ROAD LIGHTING

ROAD LIGHTING

Ir. W. J. M. van Bommel

Prof. J. B. de Boer

PHILIPS TECHNICAL LIBRARY
KLUWER TECHNISCHE BOEKEN B.V. -
DEVENTER-ANTWERPEN

Cover photograph: Whitestone Bridge, New York, lighted with 180 watt low-pressure sodium lamps (photograph by courtesy of International Lighting Review).

ISBN 90 201 1259 7

© 1980 Kluwer Technische Boeken B.V. - Deventer

1e druk 1980

Preface

This is the second book in the Philips Technical Library series to be devoted to the subject of road lighting. The first, entitled Public Lighting, appeared in 1967. Since that time, however, the art and science of road lighting have both progressed to keep pace with the latest results of the continuing research being conducted in an ever increasing number of countries throughout the world. The purpose of this present book, which represents a complete and fresh approach to the subject, is to outline the underlying principles on which modern road lighting design is based and to leave the reader with an understanding of the background to the subject and a knowledge of how these principles should be applied in practice.

The book is divided into four parts. The five chapters comprising Part One discuss the fundamentals of road lighting. Extensive reference is made to research carried out both in the lighting laboratories of Philips and elsewhere. The relevant lighting criteria are analysed from the point of view of both the visual performance and the visual comfort of the road user, and the influence here of different lamp spectra are outlined. The section concludes with a look at the fundamentals of tunnel lighting, in so far as they differ from those applicable to the lighting of open roads.

Part Two of the book describes the equipment important for road lighting. Lamps and luminaires are reviewed in terms of their practical properties and features, the intention being to facilitate a proper selection of this equipment for the various application fields rather than give the technical background to its development. These properties, together with the reflection properties of a road surface serve to determine the perceived brightness of that surface. Road surface brightness being an important quality parameter, the last chapter in the section deals with the question of how road surface reflection properties are defined and used in lighting design.

Part Three of the book provides the link between theory and practice and supplies the reader with the knowledge needed for effective lighting design. The main lighting criteria applicable to both open road and tunnel lighting are reviewed in the form of a summary of the results arrived at in Part One. This is followed by an examination of how these criteria have been incorporated into the various national and international lighting recommendations. Guidelines on how to avoid a crucial decrease in lighting quality during adverse weather conditions are given, as are the maintenance operations needed in order to prevent an intolerable deterioration of road lighting installations with time. It is, of course, of the utmost importance that lighting installations be designed to meet the required quality standard, whilst at the same time resulting in minimum costs and energy consumption. A chapter is therefore devoted to these aspects of design. The various calculation methods available for use in designing road lighting installations are also outlined, while in the last chapter of this part both laboratory measurements on lighting equipment and field measurements of the basic lighting quantities are described.

The fourth and final part of the book describes various practical solutions to lighting problems in the different fields of application. There are three chapters: Roads and Junctions, wherein practical advice is given on how to arrive at the best solution for a project based on the consideration of the lighting criteria for motorised traffic discussed earlier in the book; Residential and Pedestrian Areas, where the emphasis is on those factors that are important as seen from the point of view of local pedestrians and inhabitants of the area concerned; and finally Tunnels and Underpasses, which discusses the various alternative practical solutions in this field. In several instances, examples of lighting installations have been chosen from projects completed in the Philips Lighting Design Centres throughout the world.

This book has not been aimed solely at the public lighting engineer or the student of lighting engineering; there is much of interest here too for the traffic engineer, town architect, town planner and road construction engineer.

We thank Prof. Dr. D. Fischer (Eindhoven University of Technology) for his careful reading of the manuscript and for his many valuable comments which, we are sure, have led to important improvements and have helped to make this book part of a series which started in 1978 with the book 'Interior Lighting' (by J. B. de Boer and D. Fischer). We should also like to acknowledge the work of Mrs. A. Thomas who prepared the typescript and Mr. J. van Hemert who produced the many drawings.

Finally, we are much indebted to Mr. D. L. Parker who assisted in the preparation of the manuscript and advised us on many points.

Eindhoven, May 1980
W. J. M. van Bommel
J. B. de Boer

Contents

Part 1

FUNDAMENTALS

Chapter 1

Lighting Criteria

The safety and comfort of a road user deteriorate considerably with the onset of darkness, particularly on those roads not provided with a well designed and maintained lighting installation. With regard to safety, for example, it has been shown (Janoff, *et al.* 1977) that more than 50 per cent of all road deaths in the USA occur at night. Weighted for kilometres travelled, the night fatality rate in the USA is approximately two and a half times the day-time figure. In Great Britain, the number of night-time accidents (including non-fatal accidents), again weighted for kilometres travelled, is some 1.8 times that by day (Sabey, 1976).

Studies in many countries and by many different institutions have shown that road lighting can help to reduce the number of night-time accidents by more than 30 per cent (Fisher, 1977 and CIE, 1979c). Some of these studies are of the so-called before-and-after type, in which figures for night-time accidents before and after a change in the lighting are compared, the number of day-time accidents serving as a correction factor for possible changes in speed, traffic density and so forth, which may have taken place during the study period. The other studies are those in which night-time accident figures are compared for similar sites that differ only with regard to the lighting.

Some idea of the sort of accident reductions that have been obtained in various parts of the world and on different types of road through the provision of road lighting can be obtained by a study of table 1.1. Only those results that are statistically significant have been included.

As regards the costs of running a road lighting installation, it has long been realised that these could to a large extent be offset against the saving to the community in the cost of accidents prevented. It would appear from a recent study (Scholz, 1978) that lighting becomes economically viable if it can account for a mere 15 per cent reduction in the number of night-time road accidents, that is to say half the reduction often realised in practice.

There is thus good evidence to suggest that road lighting can reduce the number of accidents occurring at night, and that this lighting can be largely paid for out of the savings to the community that it brings with it. It would seem worthwhile, therefore, to look for a possible relationship between accident reduction figures and the standard or quality of the lighting installations producing them; such a relationship could then be used as a sound basis for the design and specification of truly effective road lighting.

Unfortunately, whilst the results of certain studies indicate that accident reduction improves with increasing lighting level (ITE/IES, 1966; Skene, 1976; and Cobb *et al.*, 1979), no significant results are available on the relationship between the incidence of road accidents and other lighting parameters, such as for example lighting uniformity and glare. And there is, of course, no question of deliberately creating a road lighting situation in which accidents are likely to occur just in order to acquire such data. It is unlikely, therefore, that a significant conclusion concerning the relation between the quality of

Table 1.1 *Percentage reductions in the number of night-time road traffic accidents resulting from the provision of road lighting*

	Country	Accident reduction (%)	Type of accident	Remarks	Reference
motorways	U.S.A.	62	injury	dual carriage-way (3-lane) 1966	ITE/IES (1966)
	U.S.A.	40 52	injury	dual carriage-way (2 and 3 lane) 1971	Box (1971)
	Japan	56	all	dual carriage-way (2-lane) 1973	Fujimori (1973)
trunk roads	Great Britain	76	injury	3-lane 1962	Christie (1962)
	Great Britain	38	injury	dual carriage-way 1966	Christie (1966)
	Great Britain	44 30 38	injury injury injury	1972	Cornwell/ Mackay (1972)
	Great Britain	53	fatal and serious injury	mainly single carriageway 1973	Sabey/Johnson (1973)
roads in built-up areas	Switzerland Australia	36 57 21 29	injury pedestrian non pedestrian all	1958 1972	Borel (1958) Turner (1972)
	Great Britain	30 33	injury injury	1972	Cornwell/ Mackay (1972)

14

road lighting and night-time accident rates will ever be arrived at.

What can and has been attempted, however, is to examine the connection known to exist between lighting quality and the visual reliability of road users. By visual reliability is meant the ability of a motorist to continuously select and process, more or less subconsciously, that part of the visual information presented to him that is necessary for the safe control of his vehicle. The reasoning behind this approach is, of course, that if visual reliability can be improved it must inevitably lessen the chances of accidents occurring.

The visual reliability of a road user is heavily dependent on his ability to detect subtle changes in his field of view; in other words, his visual performance has to be good. But for a high level of visual reliability to be maintained, especially when driving conditions are less than ideal, the road user must feel comfortable in the visual environment created by the road lighting. The first thing to determine, therefore, is precisely which are the lighting parameters that positively influence these two components of visual reliability, namely visual performance on the one hand and visual comfort on the other.

However, before discussion on these points can proceed, it is first necessary to establish which basic lighting unit plays the dominant role in the seeing process, illuminance or luminance.

1.1 Luminance and Vision

A surface is made visible by virtue of light being reflected from it and entering the eye of the observer: the greater the amount of light entering the eye, the stronger will be the visual sensation experienced. Thus, the illuminance on a road surface, which refers only to the amount of light reaching that surface per unit of area, can give no indication of how strong the visual sensation will be, or in other words how bright the surface will appear. The brightness will depend on the amount of light radiated by the surface per unit of bright area and per unit of solid angle in the direction of the observer. This is the luminance (L) of the surface, which is given by

$$L = Eq$$

where E is the illuminance on the surface and q is its luminance coefficient, which is a measure of the amount of light reflected by the surface in the direction of the observer.

That it is the luminance and not the illuminance that determines the brightness of a particular point is illustrated by way of the four photographs of one and the same road lighting installation shown in figure 1.1. The illuminance pattern on the road is the same in each photograph; it is the changes in the reflection properties of the road surface which have led in turn to changes in the luminance pattern that is responsible for the differences in brightness.

Figure 1.1 The influence of road-surface reflectance on perceived surface brightness, with illuminance level and distribution constant: (a) Smooth-dry surface; (b) Smooth-wet surface; (c) Rough-dry surface; (d) Rough-wet surface.

Since brightness is finally determined not by illuminance but by luminance, the visual performance and visual comfort of a road user are directly influenced by the complex pattern of luminances existing in his view of the road ahead. It is only to be expected, therefore, that lighting parameters based on the luminance concept will be those best suited for use in defining these two components of visual reliability.

1.2 Visual Performance

1.2.1 Lighting Level

At the relatively low lighting levels common in road lighting, colour vision is poor and visual detection is made possible more by the difference in lumi-

nance between an object and its background than by any difference in colour
An object is said to possess a luminance contrast (C) defined by

$$C = \frac{|L_o - L_b|}{L_b}$$

in which L_o is the luminance of the object itself and L_b the luminance of the
background against which it is seen.

If an object is darker than its background it will be seen in silhouette and its
contrast is said to be negative; if, on the other hand, it is brighter
than its background its luminance contrast is said to be positive. It will be
shown later that road lighting produces mainly negative contrasts.

The luminance contrast needed for an object to become visible depends,
amongst other things, on the luminances surrounding the object, as it is these
that determine the adaptation condition of the observer's eye. This is illustrated in figure 1.2, which gives the relation between the just perceivable
contrast – termed the threshold contrast, C_{th} – and background luminance for
a standard object (CIE Publication No. 19, 1972). The figure shows that C_{th}
decreases as the background luminance increases.

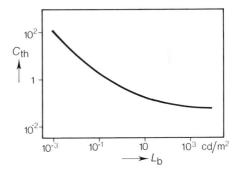

*Figure 1.2 Decrease in threshold contrast
(C_{th}) with increase in background luminance
(L_b) for a disc subtending an angle of
4 minutes of arc and presented for $1/5$ s at
1 s intervals. (CIE Publication No. 19)*

Increasing the luminance of the background against which an object is
viewed, viz. increasing the luminance of the road surface, thus increases the
chances of the object being detected, because the threshold contrast for
objects in general is decreased.

This influence that background luminance has on the detectability of an
object is illustrated in figure 1.3. The shaded area in each column defines the
range of diffuse (object) reflectance (ρ) over which an object (in fact that
defined in figure 1.2) becomes undetectable. This range is clearly smaller at
the higher background luminance (1 cd/m²) than at the lower, a fact attributable to the decrease in contrast threshold with increase in background luminance shown in figure 1.2. It is interesting to note that due to this effect, the
range of reflectances for which positive contrast seeing occurs is greater for

17

the higher background luminance, provided similar luminaires and luminaire arrangements are employed, viz. the ratio of object illuminance to background luminance (E_v/L_b) should remain constant. (This effect explains similar results found by Ketvertis, 1977 in full-scale tests.)

A further phenomenon illustrated by this figure is that a decrease in the vertical illuminance on the object raises the lower limit of object reflectance at which an object becomes undetectable. This is due, of course, to the increased negative-contrast seeing involved at the lower vertical illuminance.

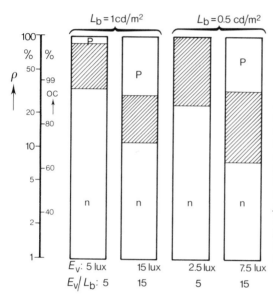

Figure 1.3 Range of diffuse reflectance (ρ) over which an object is detectable in positive (p) and negative(n) contrast, calculated with the aid of figure 1.2 for different background luminances (L_b) and different vertical illuminances (E_v). The shaded areas represent the ranges for which objects are not detectable. The scale marked OC gives the percentual cumulative occurrence of reflectances of pedestrians' clothes (Smith, 1938).

The improvement in visual performance arising from an increase in background luminance or a decrease in vertical illuminance, or both, can perhaps be better appreciated by studying the scale in figure 1.3 marked in terms of the percentual cumulative occurrence of reflectances of pedestrians' clothes (according to Smith, 1938). This scale reveals that due to an increase in L_b, the number of detectable objects likely to be present on a road increases considerably. The objects that remain undetectable under road lighting alone are those with high reflectance values, and these become easily visible in positive contrast with the aid of vehicle headlighting.

Needless to say, an object having a very high reflectance will usually be clearly visible in positive contrast, no matter how good or poor the road lighting, when lighted by even the poorest headlights. It is for this reason, of course, that pedestrians are advised to wear an item of highly reflective clothing, or better still one that works on the fluorescent (Day-glo) principle, when obliged to walk along poorly lighted or unlit roads at night.

Summarising, it has been shown that under fixed road lighting conditions, visual performance improves with increase in road surface luminance and with decrease in vertical illuminance. Provided the vertical illuminance can be controlled, therefore, the average road surface luminance (L_{av}) can be considered an important lighting parameter.

1.2.2 Uniformity

To ensure not only a good average but also a certain minimum for the visual performance at all locations on the road, the difference between the average and the minimum road surface luminance should not be too great. This can be ensured by specifying a minimum for the ratio of minimum to average road surface luminance. A second lighting parameter is therefore this luminance ratio, which is called the overall uniformity (U_o).

The smaller this ratio (i.e. the poorer the uniformity), the worse will be the visual performance for objects seen against the low-luminance part of the road surface, which is where the contrast value is low and the contrast threshold high. But large luminance differences in the field of view also result in a lowering of the contrast sensitivity of the eye, and give rise to so-called transient adaptation problems.

As Adrian and Eberbach (1968/1969) have shown, the latter effect can be explained by considering the brighter parts as forming glare sources for an observer looking towards the darker parts of the road surface.

1.2.3 Glare

The mechanism by which the loss of visual performance due to the presence of glare takes place can best be understood by considering the light scatter taking place in the eye of the observer, figure 1.4. A sharp image of the scene in the direct field of view is focussed on the retina of the eye, the resulting visual sensation being determined by the luminance of the scene. At the same time, however, light coming from a glare source that is too close to the direct line of

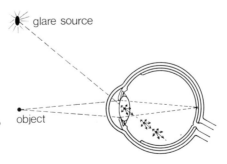

Figure 1.4 Light scatter in the eye due to glare.

sight is partly scattered in the eye, and that part of it falling unfocussed on the focussed image produces a visual sensation that can be likened to the drawing of a bright veil across the field of vision.

This veil can be considered as having a luminance – the equivalent veiling luminance – proportional to the degree of scatter in the direction of the retina. The overall strength of the visual sensation is then determined by the sum of the two components: scene luminance and equivalent veiling luminance. Holladay (1927) found the latter component to be dependent upon the illuminance (E_{eye}) on the eye and the angle (θ) between the viewing direction and the direction of light incidence from the glare source. A number of workers (Fisher and Christie, 1965; Hartmann and Moser, 1968; and Adrian, 1975) have shown that for the range of luminances normally occurring in road lighting and for θ within the range 1.5° to 60°, veiling luminance can be given by the empirical formula

$$L_v = 10\,\frac{E_{eye}}{\theta^2}$$

where L_v = equivalent veiling luminance (cd/m²)
$\quad\quad\ E_{eye}$ = illuminance on the eye produced by the glare source, in a plane
$\quad\quad\quad\quad\quad$ perpendicular to the line of sight (lux)
$\quad\quad\ \theta$ = angle between direction of view and direction of light in-
$\quad\quad\quad\quad$ cidence from glare source (degrees)

The value of 10 for the constant of proportionality (with the dimensions degrees²/steradian) is representative for observers aged between 20 and 30 years. Fisher and Christie (1965) have shown that this value increases with the age of the observer at the rate of roughly 0.2 degrees²/steradian per year. The limits imposed on θ will be met so long as the observer views only the road surface straight ahead, and provided the screening angle of the car roof is taken into account, the latter being standardised by the CIE (1976b) at 20°. Crawford (1936) showed that for multiple glare sources the total equivalent veiling luminance could be described by adding the equivalent veiling luminances of the individual sources, thus

$$L_v = \sum_{i=1}^{n} L_{vi}$$

The overall effect of glare on visual performance can now be determined by first adding the total equivalent veiling luminance to each of the two luminances $(L_o$ and $L_b)$ forming the contrast of the task object. The result of this is twofold

20

(1) the contrast threshold of the task object decreases due to the increase in effective background luminance, i.e.

$$L_b \rightarrow L_{b\ eff} = L_b + L_v$$

(2) the contrast decreases from

$$C_0 = \frac{|L_o - L_b|}{L_b} \text{ to } C_{eff} = \frac{|(L_o + L_v) - (L_b + L_v)|}{L_b + L_v} = \frac{|L_o - L_b|}{L_b + L_v}$$

i.e. $C_{eff} = \dfrac{L_b}{L_b + L_v} C_0$

With the contrast threshold curve given in figure 1.2 it can be proved that, as might be expected, the positive effect of the decrease in contrast threshold is not sufficient to compensate for the loss of effective contrast. In other words, an object that can just be seen when there is no glare (threshold contrast) cannot be seen when glare is present unless the actual contrast is increased. This effect forms the basis of the measure for loss of visual performance due to glare, the so-called threshold increment. This is defined as the amount of extra contrast required to again just make the object visible under glare conditions, divided by the effective contrast. (The same value of threshold increment will be obtained if the increment in luminance difference between object and background needed to see the object again under glare conditions is divided by the threshold luminance difference without glare.)

The CIE (1976b) prescribes that the threshold increment should be determined on the basis of the perception of an object that subtends an angle of 8 minutes of arc at the eye of the observer, and refers to the threshold contrast values published by Adrian (1969), which are based on measurements made by Blackwell (1946). The threshold contrast curve for an 8-minute object is given in figure 1.5.

By way of example, the threshold increment will be determined with the aid of this curve for a situation in which the background luminance L_{b_1} equals 1 cd/m^2 and the equivalent veiling luminance L_v equals 2 cd/m^2.

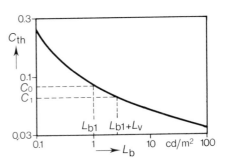

Figure 1.5 Threshold contrast (C_{th}) as a function of background luminance (L_b) for an object subtending an angle of 8 minutes of arc. (Adrian) The threshold contrast (C_o) prevailing at $L_b = L_{b1} = 1\ cd/m^2$ decreases to C_1 when the veiling luminance $L_v = 2\ cd/m^2$.

The threshold contrast without glare can be read off from the curve at $L_{b_1} = 1$ cd/m², viz. $C_o = 0.081$. Add to this value of L_b the specified veiling luminance of 2 cd/m², and the new contrast threshold becomes $C_1 = 0.061$. The effective contrast, C_{eff}, is then given by

$$C_{eff} = \frac{L_{b_1}}{L_{b_1} + L_v} C_o = 0.081/3 = 0.027$$

To make the object visible again with this amount of glare, the contrast C_{eff} would have to be increased to C_1. The threshold increment, in per cent, is therefore

$$\frac{C_1 - C_{eff}}{C_{eff}} 100 = \frac{0.061 - 0.027}{0.027} 100 = 126 \text{ per cent}$$

In practice, of course, contrasts will not be increased, and the value of the threshold increment then indicates the loss of visual performance due to glare. For each different background luminance, that is to say for each part of the road surface viewed by an observer, the effect of glare on visual performance will be different, and a different value for the threshold increment will be obtained. The CIE (1976b) therefore recommends employing as an overall measure of disability glare (the glare form associated with a deterioration of visual performance) the relative threshold increment (TI) based on the average road surface luminance value as background luminance. The CIE states that for the luminance range 0.05 cd/m² < L_b < 5 cd/m² the threshold increment is approximated with the formula

$$TI = 65 \frac{L_v}{L_{av}^{0.8}}$$

where TI = relative threshold increment (per cent)
        L_v = equivalent veiling luminance (cd/m²) for an observer looking straight ahead in a direction parallel to the road axis and 1° down from the horizontal
        L_{av} = average road surface luminance (cd/m²)

A third road lighting parameter is thus the threshold increment, TI.
It has been shown that visual performance can also be improved by limiting the component of the lighting responsible for the vertical illuminance. This limitation is automatic in those symmetrical luminaires in which glare restriction measures have been incorporated in their design, their construction then being such that the luminous intensities at large angles from the downward vertical are insufficient to produce dangerously high vertical illuminances. There is therefore usually no need to specify the vertical illuminance, provided the glare restriction specification is good.

Summarising, three basic road lighting parameters have been found each of which influences the degree of visual performance of the road user. These parameters are
(1) The average road surface luminance L_{av} (level)
(2) The overall uniformity ratio U_0 (uniformity)
(3) The threshold increment TI (disability glare)
In Chapter 2, the influence that each of these parameters has on visual performance will be discussed in detail.

1.3 Visual Comfort

The need for visual comfort on the part of road users may appear, at first, to be a luxury requirement. But, as was mentioned earlier, visual comfort is in fact most essential to road safety, since together with visual performance it sets the level of visual reliability attainable.

1.3.1 Lighting Level

It goes without saying that the degree of visual comfort experienced by a road user is very much dependent upon the value of the average road surface luminance to which he is adapted; the higher this luminance (so long as it remains below the glare level), the more comfortable he will feel in the performance of the driving task. But adaptation is in turn strongly influenced by the luminances appearing in the main field of view, viz. on the road ahead. Thus, a lighting parameter of major importance in connection with the visual comfort aspect of road lighting is the average road surface luminance (L_{av}); the same lighting parameter that was shown earlier to be important as regards visual performance.

1.3.2 Uniformity

Visual comfort is also known to be strongly influenced by the inherent unevenness of road lighting. The sort of unevenness that results in a continuously repeated sequence of alternate bright and dark transverse bands on the road surface can be particularly disturbing to a passing driver. This 'zebra' effect, to use a popular description, can of course be softened by reducing the luminance difference between successive bands of high and low brightness. The lighting parameter used to describe the severity of this effect is the so-called longitudinal uniformity (U_l), which is defined as the ratio of the minimum to the maximum road surface luminance on a line parallel to the axis of the road and passing through the observer position.
However, whilst U_l adequately describes the lengthwise uniformity of the

luminance pattern it says nothing regarding the luminance gradient (viz. the rate of change of luminance with distance travelled), and since this feature of the luminance pattern is also known to influence the visual comfort of a road user, a further lighting parameter has been proposed, the relative maximum luminance slope, S_{max} (De Boer and Knudsen, 1963).

This parameter is defined as the maximum luminance variation (ΔL_{max}) found over any 1 metre distance measured transversely or 3 metres measured longitudinally, expressed as a percentage of the average road surface luminance (L_{av}).

Unfortunately, whilst S_{max} is suitable for assessing the degree of visual discomfort associated with a particular luminance gradient (see Sec. 3.2) it is a difficult to measure parameter. There is, however, a reasonably good correlation between S_{max} and the longitudinal uniformity U_l described above, provided that is that the luminaire spacing is not smaller than about 30 metres. Since U_l is both easier to calculate and measure than S_{max}, the former is the parameter employed in road lighting specifications.

1.3.3 Glare

Glare has, of course, a disturbing effect on the visual comfort of a driver. The degree of visual discomfort experienced is dependent upon the design of the luminaire used in the road lighting installation and upon the design of the installation as a whole.

From many investigations (Hopkinson, 1940; De Boer, 1951 and 1967a; De Boer and Schreuder, 1967; Adrian, 1966; Adrian and Schreuder, 1970; Cornwell, 1971; and De Grijs, 1972), in which observers were asked to appraise the degree of glare experienced under both static and dynamic observation conditions, using both a model and actual installations, those characteristics of design influencing the comfort glare sensation have been determined. (Figure 1.6 shows the semi-dynamic road lighting simulator used by De Boer.)

From these investigations it was found that the degree of discomfort glare given by a road lighting installation is influenced by the following factors:

(1) Luminaire characteristics

I_{80} absolute luminous intensity (cd) at an angle of 80° to the downward vertical in the vertical plane parallel to the road axis

I_{80}/I_{88} ratio of the luminous intensities at 80° and 88° to the downward vertical in the vertical plane parallel to the road axis

F apparent flashed (light-emitting) area of the luminaires (m²) as seen at an angle of 76° to the downward vertical

C colour factor according to the spectral distribution of lamp used (see Sec. 4.2.3)

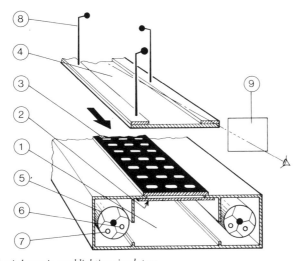

Figure 1.6 Semi-dynamic road lighting simulator:
(a) Sketch showing the main working parts. Light from the light-box (1) shines through the clear-glass plate (2) and the openings in the endless moving belt (3) to strike the underside of the frosted-glass 'road surface' (4), the luminance pattern on which can be varied by opening or closing openings in the belt. The light-box is equipped with sodium and mercury sources, (6) and (7), the selection of which is done by rotating the housing (5). Small incandescent 'luminaires' (8) move towards the observer at the same speed as the belt to create the illusion of movement along the road. The intensity and light distribution of the lanterns can be adjusted using a voltage regulator. In some experiments a neutral filter (9) was used to change the apparent brightness of the road surface.

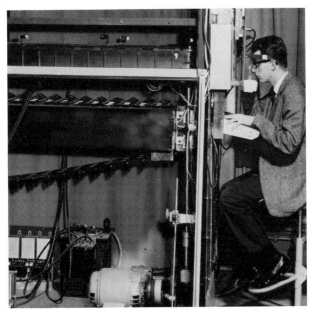

(b) The observer's seat and part of the simulator. Part of the endless belt and the light-box can be seen.

(c) Road as seen by observer.

(d) Groups of light sources used to study the influence of the source size upon the degree of discomfort glare.

(2) Installation characteristics

  L_{av} average road surface luminance (cd/m^2)
  h' vertical distance between eye level and the luminaire (m)
  p number of luminaires per kilometre

The interrelationship between these characteristics has been investigated, the aim being to describe with one figure the degree of control possessed by the installation with regard to discomfort glare. The formula accepted by the CIE (1976b) is

$$G = 13.84 - 3.31 \log I_{80} + 1.3 (\log I_{80}/I_{88})^{0.5} - 0.08 \log I_{80}/I_{88}$$
$$+ 1.29 \log F + C + 0.97 \log L_{av} + 4.41 \log h' - 1.46 \log p$$
$$= SLI + 0.97 \log L_{av} + 4.41 \log h' - 1.46 \log p$$

G being the glare control mark and SLI the specific luminaire index (see Sec. 7.2.1). It follows, of course, that since G is a glare control mark, the higher its value the greater will be the freedom from discomfort glare.
The formula for G is valid for the following ranges of values

$50 < I_{80} < 7000$ (cd) $5 < h' < 20$ (m)
$1 < I_{80}/I_{88} < 50$ $20 < p < 100$
$0.007 < F < 0.4$ (m^2)   luminaire rows $= 1$ or 2
$0.3 < L_{av} < 7$ (cd/m^2)

The influence that the important characteristics I_{80} and L_{av} have on G is shown in figure 1.7 for a 'standardised, single-sided installation'. The influence of the characteristics h', p, F and I_{80}/I_{88} on G is shown in figure 1.8.

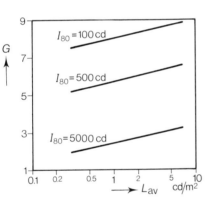

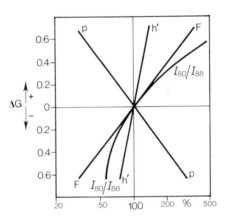

Figure 1.7 Glare control mark (G) as a function of average road-surface luminance (L_{av}), with luminaire luminous intensity (I_{80}) as parameter, for a single-sided installation in which h' = 7.5 m, p = 33, F = 0.07 m^2 and I_{80}/I_{88} = 1.75.

Figure 1.8 Incremental change in glare control mark (ΔG) with percentage change in the luminaire characteristics I_{80}/I_{88} and F, and the installation characteristics h' and p, defined in the text.

The amount of glare experienced by a road user does not remain constant, but is continuously varying by virtue of his forward movement along the lighted road. This variation in glare can itself give rise to a feeling of discomfort. The effect can be described by the variation in the threshold increment TI as obtained from measurement or calculation for different points along the road. However, since the correlation between this variation and the longitu-

27

dinal uniformity is generally good (large variation, low longitudinal uniformity), it is usually enough to specify only the latter.

To summarise, three basic lighting parameters have been introduced for
describing the visual comfort aspect of road lighting installations. These are
(1) The average road surface luminance L_{av} (level)
(2) The longitudinal uniformity U_l (uniformity)
(3) The glare control mark G (discomfort glare)
In Chapter 3, the influence that each of these parameters has on visual
comfort will be discussed in detail.

Chapter 2

Visual Performance

A motorist's visual performance can be assessed using one or more of a number of different performance criteria such as visibility distance; detection probability; revealing power; supra-threshold visibility; reaction performance, and detection of relative movement. This being the case, it might at first sight seem perfectly feasible to specify road lighting indirectly by way of these performance criteria. Although such an approach is indeed sometimes proposed, it is not to be recommended. It by no means follows that lighting chosen to enhance one aspect of visual performance will prove equally effective with regard to visual performance in general, and it is for this reason that the authors of this book advocate specifying road lighting in terms of the various lighting parameters direct. Such an approach is only possible, of course, if one can define the influence that those lighting parameters have on visual performance in general. Fortunately, there are many recent research results to draw on and this means that useful relationships between the most important performance criteria and the various lighting parameters established in the previous chapter can be fairly easily arrived at.

Before considering the first of the six performance criteria covered in this chapter, it is necessary to say something about the various target objects employed in this sort of research. These usually vary in height between about 12 cm and 20 cm, which corresponds to an angle subtended at the eye of the observer of between 4 and 7 minutes of arc when viewed from a distance of 100 m. An object measuring 20 cm × 20 cm is often adopted as a suitable 'critical obstacle' in tests calling for avoiding action on the part of the motorist, the clearance between the underside of most cars and the road surface generally being of this order.

2.1 Visibility Distance

An obvious measure of the visual performance obtainable under defined lighting conditions is the distance at which a given obstruction on the road is visible.

This distance, termed the visibility distance, has been determined by De Boer, et al. (1959) for a vertically placed 20 cm × 20 cm matt screen having a reflection factor of 9 per cent approached at a speed of 50 km/h on a number of differently lighted roads. From the results of their research these authors concluded that in order to detect the screen at the safe stopping distance of 100 metres when its contrast was 0.30 called for an average road surface luminance of 2.2 cd/m².

A more recent study along these lines has been conducted by Economopoulos (1978). He measured the visibility distance at two driving speeds and at a number of luminance levels on lighted roads outside built-up areas; roads that were closed to normal traffic. The driver-observers were instructed to look out for a series of three obstacles along a five-kilometre stretch of road,

a

b

Figure 2.1 The obstacle placed for observation at a point on the road where the contrast is (a) a minimum and (b) a maximum. The road lighting is of the 'opposite' arrangement type. (Economopoulos)

30

each obstacle consisting of a vertically placed matt plate measuring 20 cm × 20 cm having a reflection factor of 11 per cent. Unknown to the driver-observers, the obstacles were placed at those positions on the road where the contrast was a minimum, a maximum and average (figure 2.1).

The different lighting (luminance) levels were obtained by having the observers wear neutral spectacles with differing transmission factors. The road lighting installations were equipped with low-pressure sodium lamps mounted in luminaires defined by CIE as being of the tight-glare-control variety. Results are given in figure 2.2, where the average visibility distance for two driving speeds (40 km/h and 80 km/h), has been plotted as a function of the average road surface luminance, with object contrast as parameter.

Knowing the visibility distance, the reaction time of a driver and the deceleration rate of the vehicle, it is possible to determine the safe speed limit for the conditions prevailing. Figure 2.2 can then be transformed to give the relation between safe driving speed and average road surface luminance. Economopoulos did this taking, by way of example, a reaction time of 2 seconds and a deceleration rate of 4 m/s^2. His results are shown in figure 2.3 for two values of contrast.

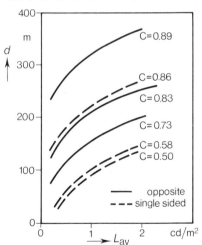

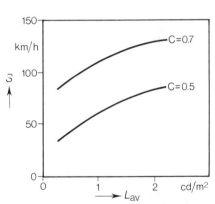

Figure 2.2 Increase in visibility distance (d) with increase in average road-surface luminance (L_{av}) for two types of installation, with contrast (C) of 20 cm square obstacle (reflection factor 11 per cent) as parameter. (Economopoulos)

Figure 2.3 Safe driving speed (S) (based on a driver reaction time of 2 s and a deceleration rate of 4 m/s^2) as a function of the average road-surface luminance (L_{av}) for two values of object contrast (C). Based on the visibility distances given in figure 2.2. (Economopoulos)

It would appear that the safe speed limit for a road surface luminance of, say, 2 cd/m^2 would be a little under 90 km/h in the case of the object having a contrast $C = 0.5$ and about 120 km/h for the object where $C = 0.7$. However, the fact that the observers employed in this work were alerted to the presence of obstacles in the road means that it is only to be expected that had they not been so alerted, somewhat shorter visibility distances would have resulted. In practice, therefore, the safe driving speeds would be somewhat lower than those indicated above.

The object contrasts met with in practice are fixed by a number of factors: namely, the reflectance of the object itself, the luminance of the road surface against which it is viewed and the vertical illuminance on the object. Contrasts lower than 0.5 will often be encountered under road lighting conditions.

2.2 Detection Probability

Narisada (1971), using a 1/25-scale lighting simulator, investigated the influence of road surface overall uniformity and average luminance upon the probability of a motorist detecting a defined object in his path. Observers, ranging in age from 16 to 30 years, were asked to detect the presence of obstacles, each with an apparent size of 20 cm $\times$ 20 cm, located against the darkest part of the road surface. The road surface was lit by tight-glare-control light boxes. The obstacles always formed a minimum luminance contrast of about 0.25 with their background and were exposed for 0.5 seconds during each observation at apparent distances of 60 m and 100 m in front of the observer.

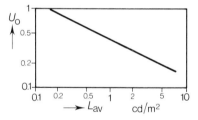

Figure 2.4 Relationship between average road-surface luminance (L_{av}) and overall uniformity (U_o) needed for a 75 per cent probability of detecting a 20 cm square object seen against the darkest part of the road. (Narisada)

Figure 2.4 shows the relation between the average road surface luminance (L_{av}) and the overall uniformity (U_o) needed to achieve a detection probability of 75 per cent. It can be seen that this degree of probability will be obtained when, for example, $U_o = 0.4$ and L_{av} is about 1.5 cd/m^2; further, that halving the uniformity will call for a fourfold increase in L_{av}.

2.3 Revealing Power

A method of using given values of the lighting parameters to calculate the number of objects out of a defined set of objects detectable at various points on a road was first described over forty years ago by Waldram (1938). He used as a basis for his calculations a relation known to exist between the contrast

32

threshold and the background luminance (a relation similar to the one recently published by the CIE (1972) and shown earlier in figure 1.2). The set of objects was defined by the curve representing the probability of the occurrence of a reflection factor of pedestrian clothing not exceeding some given value, as measured by Smith (1938) and shown in figure 2.5. (A far more recent investigation has shown only slight changes in the trend of this curve, Van Bommel, 1970.)

Waldram called the percentage of objects detectable at each point on the road the revealing power of the lighting, a concept that was later to be used by other workers. (Harris and Christie, 1951; Hentschel, 1971; Lambert, *et al.* 1973.)

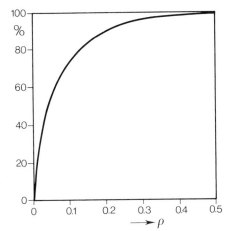

Figure 2.5 Percentage of persons wearing clothing having a reflectance (ρ) not exceeding a given value. (Smith)

In order to demonstrate the influence that the lighting parameters level (L_{av}), uniformity (U_o) and glare (TI), introduced in the previous chapter, have on visual performance Van Bommel (1979) has used Waldram's method to calculate the revealing power for the darkest position on the road (local minimum road surface luminance) under various road lighting conditions, each of which is characterised by one of three combinations of U_o and TI.

The calculations, which were repeated for various values of the vertical illuminance and average road surface luminance, are based on the threshold contrast curve and on the reflectance distribution curve mentioned above.

The results of these calculations are presented graphically in figure 2.6 by means of lines of constant revealing power. Consider first the curves of figure 2.6a, which is valid for tight-glare-control luminaires resulting in a TI value of 7 per cent and a U_o value of 0.4. For a normal type of asphalt road surface at the point where the road surface luminance is a minimum, such an installation will produce a vertical illuminance (in lux) that is numerically equal to about five times the average luminance (in cd/m²): thus, for $L_{av} = 2$ cd/m² E_v will be approximately 10 lux at the point where the road surface luminance is min-

33

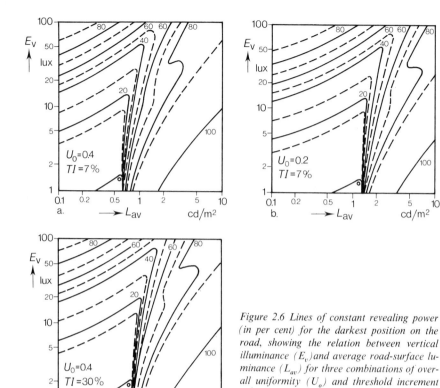

Figure 2.6 Lines of constant revealing power (in per cent) for the darkest position on the road, showing the relation between vertical illuminance (E_v) and average road-surface luminance (L_{av}) for three combinations of overall uniformity (U_o) and threshold increment (TI): (a) $U_o = 0.4$, TI = 7% (b) $U_o = 0.2$, TI = 7% (c) $U_o = 0.4$, TI = 30%.

imum. The revealing power is then about 85 per cent: this means that under the conditions specified, 85 per cent of the objects can be seen. Suppose now that the lumen output of the lamp were reduced to give a lower value of L_{av}, then the vertical illuminance would be reduced in the same proportion; for example, at $L_{av} = 1$ cd/m² E_v would be reduced to 5 lux, and the revealing power would be 70 per cent instead of the former 85 per cent. Similarly, at 0.5 cd/m² and 2.5 lux the revealing power decreases to well below 10 per cent. The influence that the uniformity has upon reliability of perception becomes evident when comparing figure 2.6a with figure 2.6b, which is calculated for a lower degree of uniformity (U_o equals 0.2 instead of 0.4) but with TI held constant at 7 per cent. The revealing power at 2 cd/m² has decreased from 85 per cent to 55 per cent. Only by increasing L_{av} to 9 cd/m² ($E_v = 45$ lux) can the original 85 per cent revealing power be restored.

Increasing the amount of disability glare results in the situation shown in figure 2.6c. Less tightly controlled luminaires have been employed to bring

34

about an increase in *TI* from 7 to 30 per cent, which has also resulted in a higher $E_v : L_{av}$ ratio of 7.5 : 1 instead of the previous 5 : 1. Thus, at $L_{av} = 2$ cd/m² ($E_v = 15$ lux), because of the increased glare, the revealing power has dropped to 70 per cent.

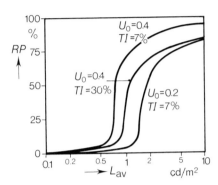

Figure 2.7 Revealing power (RP) at the darkest location on the road, as a function of the average road-surface luminance (L_{av}) for three combinations of overall uniformity (U_o) and threshold increment (TI). For TI = 7% and 30%, the vertical illuminance (in lux) is numerically equal to 5 and 7.5 times the average road-surface luminance (in cd/m²), respectively.

The results derived from figure 2.6 are depicted by the three curves of figure 2.7 in which revealing power has been plotted as a function of average road surface luminance. It is interesting to note that a 75 per cent revealing power is obtained for the condition $U_o = 0.4$ and $L_{av} = 1.1$ cd/m² but also for $U_o = 0.2$ and $L_{av} = 3.9$ cd/m². In other words, to maintain the revealing power at this level whilst halving the degree of uniformity the average road surface luminance must be increased by a factor of about 3.5. This is close to the factor of 4 found by Narisada in his work on detection probability mentioned earlier (see figure 2.4). Equal 75 per cent revealing power is also obtained for the condition $U_o = 0.4$ and $TI = 30$ per cent, but only by increasing L_{av} from 1.1 cd/m² to 2.7 cd/m², a factor increase of about 2.5

2.4 Supra-threshold Visibility

For an object to be visible at all it must, of course, be at or above the so-called threshold of visibility. The visual performance of a driver will only be adequate from the traffic safety point of view if he is able to detect obstacles in his path with the minimum of effort. This will only be the case under supra-threshold conditions. The supra-threshold level of visibility can be expressed in terms of the visibility level attainable, or it can be awarded a visibility index.

2.4.1 Visibility Level

Visibility level, the term introduced in CIE Report 19 (1972) for describing visibility in supra-threshold conditions, is defined as the ratio of the actual luminance contrast (*C*) of the CIE reference task (4 minute disc) presented under CIE reference conditions (see caption to figure 1.1) to the luminance contrast (C_{th}) of this task in the threshold situation at the same luminance of the surround

$$VL = C/C_{th}$$

In practice, however, the visual tasks that must be performed by a motorist bear little resemblance to the CIE reference task, either as regards the size and shape of the task objects or the criteria used to measure task performance. The concept of an equivalent task contrast ($\tilde{C}$) has therefore been introduced to serve as a measure of difficulty for a given task under given conditions of lighting, where

$$\tilde{C} = rC_{th}$$

r being the factor by which the contrast of the given task must be reduced to bring it to the threshold of visibility and C_{th} the contrast threshold of the CIE reference task under the lighting conditions considered.

The visibility level for the arbitrary task is then defined by

$$VL = \tilde{C}/C_{th} = r$$

The visibility level for this task is thus known once the factor r has been determined, and this can be done using a so-called contrast reducing visibility meter, an instrument developed by Blackwell (1970) specifically for this purpose. (Alternatively, the visibility level can be approximated by calculating $\tilde{C}$ from the task contrast as described by Blackwell and Blackwell (1977); according to these workers the results obtained using their calculation method are quite satisfactory for design purposes.)

It is known from the results of investigations based on typical office tasks that there is a relation between task performance – defined as the correctness and speed with which a task of given $\tilde{C}$ can be detected – and visibility level.

It should be quite clearly pointed out that similar relations applicable to the sorts of tasks met with in practice by the motorist have not yet been established.

2.4.2 Visibility Index

An alternative measure of supra-threshold visibility, developed specially for evaluations in road lighting, and which is simpler than visibility level in that it avoids the use of a visibility meter, is the so-called visibility index (*VI*) proposed by Gallagher and Meguire (1975) where

$$VI = (C/C_{th})DGF$$

in which C = task contrast under the road lighting conditions actually
prevailing

C_{th} = threshold contrast of CIE reference task at the background luminance, L_b

DGF = disability glare factor

The disability glare factor, which serves to compensate for the loss of visibility
due to glare, is in turn given by

$$DGF = \frac{L_b \, C_{th}}{L_{bg} \, C_{thg}}$$

where L_b = normal background luminance

L_{bg} = effective background luminance with glare present

$$L_{bg} = \frac{L_b + L_v}{1.074} \text{ in which } L_v = \text{equivalent veiling luminance}$$

C_{th} = normal contrast threshold

C_{thg} = contrast threshold at L_{bg}

Thus, unlike VL, VI can be evaluated direct from the normal lighting
parameters.

Gallagher (1976) has shown that this index is directly related to driver performance on city streets. Janoff, *et al.* (1978) extended this relationship to actual
night-time accident experience. According to Janoff there exists a certain
correlation between the so-called 15th-percentile Visibility Index of an installation (the minimum VI value occurring over 85 per cent of the installation
concerned) and the number of accidents, for a given population density and
type of area. The correlation is, however, relatively poor (correlation coefficient smaller than 0.45).

At the time of writing, this work by Janoff represents the only study of the
connection between road accident figures and a measure of visual performance. Similar relations can be expected to exist between accident rates and
the various other performance criteria employed in road lighting.

Gallagher (1976) established a relation between the visibility index and the
performance of motorists in a study of the 'target avoidance manoeuvre' of
over 1 300 unalerted motorists on a road where the lighting parameters – and
thus the visibility indices – were known. The target used was three dimensional and consisted of a 17.8 cm diameter cylinder topped by a hemisphere of
the same diameter to give a total target height of 35.6 cm. The target reflectance was 18 per cent. The performance criterion employed was the time
separation between vehicle and target at the moment an avoiding manoeuvre
was started – the so-called 'time to target'. The relation between time to target
and target visibility index is shown in figure 2.8.

Economopoulos (1978) calculated the visibility distance, from the time to target defined by Gallagher, for a speed of 40 km/h and a reaction time – time between seeing and reacting to the presence of the obstacle – of 2 seconds. The results of these calculations, together with Economopoulos's own directly measured visibility distances (as described in Sec. 2.1), are given in figure 2.9 as a function of the visibility index.

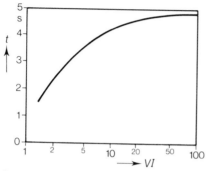

Figure 2.8 Motorists' performance represented by time-to-target (t) at the moment of starting an avoiding manoeuvre, as a function of visibility index (VI). (Gallagher)

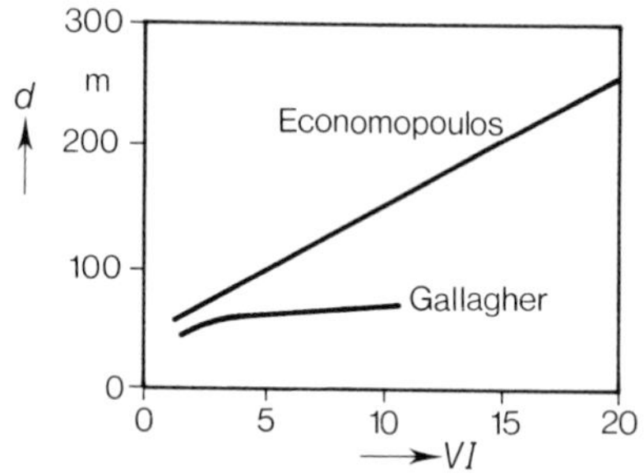

Figure 2.9 Visibility distance (d), both measured (Economopoulos) and calculated (Gallagher), as a function of visibility index (VI).

It can be seen at a glance that Economopoulos's visibility distances are far greater than those derived from Gallagher's times to target. The reason for this difference is that Gallagher did not in fact determine the visibility distance (viz. the distance between observer and target at the moment when the latter first became visible), but the reaction distance, that is to say the distance at which a first reaction to the presence of the obstacle was recorded. This reaction distance is, of course, often less than the visibility distance defined above.

The divergence of the curves in fact bears out what is so often experienced in practice: namely, that when an object is barely visible (low VI), the visibility distance and the reaction distance are likely to be equal, but that as object visibility improves there is an ever-increasing conscious delay between the moment of its perception and the moment when the avoiding manoeuvre, where necessary, is begun.

The important influence that the basic lighting parameters L_{av}, U_o and TI have on visibility index can perhaps best be demonstrated by calculating the index for an object located at the darkest point on a road for various combinations of these parameters.

This has been done (Van Bommel, 1979) for an object reflectance of 18 per cent (the reflectance employed by Gallagher) and the results, given in figure 2.10, show once again the important influence that these three parameters have on visual performance; although it should be mentioned that the improvement in VI with increase in L_{av} does in fact reach a point of saturation when the latter is increased much in excess of 10 cd/m². Nevertheless, the point to be made is that the supra-threshold criterion of visibility (VI) is still increasing at this value of L_{av} whereas the threshold criterion (RP) levels-off at this point (see figure 2.7).

Visibility index can be seen in terms of the perhaps more meaningful visibility distance (d) using the second vertical scale of figure 2.10, which has been derived from figure 2.9 (results of Economopoulos).

Figure 2.10 Visibility index (VI) at the darkest position on the road as a function of the average road-surface luminance (L_{av}) for three combinations of overall uniformity (U_o) and threshold increment (TI). For TI = 7 % and 30 %, the vertical illuminance (in lux) is numerically equal to 5 and 7.5 times L_{av} (in cd/m²), respectively. The object reflectance is 18 %. The vertical scale on the right relates VI with visibility distance (d) according to Economopoulos (figure 2.9).

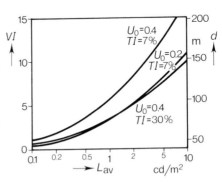

Again, as was done with revealing power, it is informative to examine the increase in L_{av} needed to maintain VI at a given level as the uniformity (U_o) and the threshold increment (TI) are in turn degraded. For the purposes of comparison, the value of VI at L_{av} = 1.1 cd/m², U_o = 0.4 and TI = 7 % (viz. VI = 6) has been taken as a starting point, the L_{av} value chosen corresponding to RP = 75 per cent in figure 2.7. Thus, from figure 2.10, L_{av} must be increased from 1.1 cd/m² to 2.6 cd/m² in order to maintain VI at 6 (or visibility distance at about 100) as U_o falls from 0.4 to 0.2, and from 1.1 cd/m² to 2.9 cd/m² to maintain this same value of VI as TI increases from 7 to 30 per cent, the factor increases in L_{av} being 2.4 and 2.6 respectively. These factors compare with the corresponding factors derived from figure 2.7 of 3.5 and 2.5 respectively.

The important influence that average road surface luminance has upon supra-threshold visibility, illustrated in figure 2.10, has been confirmed by the results of field tests carried out by Economopoulos (1977 and 1978). He extended his investigations on visibility distance, described earlier (see Sec. 2.1), by having both the visibility index and the visibility distance measured by one and the same group of observers. To this end, he built a simple device which he called a visual task analyser. With this he could determine the threshold contrast required to make the target object visible to an observer stationed 100 metres away. The ratio of the contrast actually prevailing during the driving tests to the threshold contrast determined with the visual task analyser is denoted here by the symbol VI_E, the visibility index after Economopoulos.

The confirmation can be seen from comparing figure 2.10 with figure 2.11, which shows the results of the VI_E field measurements plotted as a function of the average road surface luminance (L_{av}).

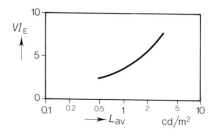

Figure 2.11 Visibility index (VI_E), as measured by Economopoulos, as a function of the average road-surface luminance (L_{av}).

2.5 Reaction Performance

As described in the foregoing section of this chapter, Gallagher and Meguire recorded the reaction of unalerted drivers to the presence of obstacles in the road. They clearly demonstrated that, up to a certain point, reaction performance improves with increasing values of the three basic lighting parameters influencing reliability of perception. However, as these workers did their observations on a relatively short stretch of road the possible influence that the lighting parameters associated with visual comfort might have on driver reaction time could not be studied.

Such a study was undertaken by Walthert (1973 and 1975). Using a static model, he studied the reaction performance of observers as certain aspects of the lighting were varied. He simplified the road luminance pattern as consisting of dark and bright transverse bars only; in this way the longitudinal uniformity, known to influence visual comfort, could be varied by altering the luminance ratio L_{min}/L_{max} of the bars. The influence of luminaire spacing was studied by varying the number of bars presented to the observer.

The reaction performance was measured as the time between the presentation of two 8-minute objects within the bars pattern and the moment when the observer signalled his detection of these.

Figure 2.12 shows the relative reaction performance (R) for objects having a contrast of 30 per cent with their background. The reaction performance clearly improves with increasing average road surface luminance, but more for low levels than for high. With decreasing longitudinal uniformity the reaction performance falls off significantly, and the closer the spacing the lower is the performance for all values of uniformity. This last-mentioned effect can be attributed to the higher luminance gradient obtained with the more closely spaced luminaires.

It must be emphasised that the results given in figure 2.12 are based on static tests. It is quite likely that under dynamic conditions the influence of both longitudinal uniformity and spacing on reaction performance would be even more pronounced.

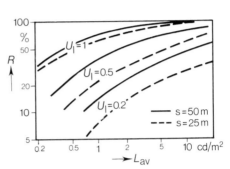

Figure 2.12 Relative reaction performance (R) as a function of the average road-surface luminance (L_{av}) for objects having a contrast of 30% with their background. Parameters are the longitudinal uniformity (U_l) of the luminance pattern (bright and dark bars) and the simulated luminaire spacing (s). (Walthert)

2.6 Detection of Relative Movement

For a driver to maintain safe control of his vehicle it is necessary not only that he should be able to see any object in his path, but that any relative movement between these objects and himself should be immediately detected. Position keeping in the traffic stream is an important aspect of the driving task, and detection of change of the angular size of the rear of the vehicle being followed is the important visual element of this task.

Fisher and Hall (1976), in a laboratory simulation, studied the time taken to react to a change in this visual angle. A square unstructured test object representing the rear view of a vehicle was projected on a screen of uniform luminance. Both the luminance of the screen and the contrast of the test object could be varied. The initial distance between the observer and the rear of the lead vehicle was simulated by the initial angular size of the test object. Acceleration or deceleration of the lead vehicle was simulated by changing the angular size of the test object by means of a zoom lens.

The measure of response to a change in visual angle was the time to unequivocal detection of change in apparent size of the test object. The observers were asked to press a button as soon as they were certain that a change in a particular direction was taking place.

The results are given in figure 2.13 for an initial distance between observer and lead vehicle of 40 metres and for two deceleration rates. It can be seen that for low values of L_{av} the detection time (t) is relatively large and decreases rapidly as L_{av} is increased.

If the difficulty of the task is increased (viz. if the deceleration rate of the lead

41

car is decreased), the detection time at a given value of L_{av} will be that much larger, although the same systematic decrease of t with increasing luminance level is found, both the curves levelling out as L_{av} approaches 10 cd/m².

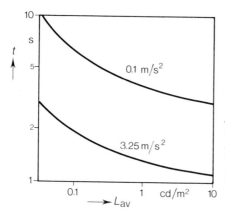

Figure 2.13 Time (t) taken to detect change in visual angle of a lead vehicle plotted as a function of the average road-surface luminance (L_{av}), with deceleration rate of lead vehicle as parameter. Contrast between lead vehicle and background was 13%. The curves are based on the results of 12 tests, two carried out by each of six observers. (Fisher and Hall)

2.7 Summing Up

Six criteria have been used to assess visual performance. In each case, the way in which the criterion concerned is influenced by the various quality parameters used in road lighting design has been discussed.

Because of the almost infinite variety of visual tasks occurring in practice, it is important that road lighting recommendations be based on general conclusions arrived at from a study of all these relations rather than on information gained from any one relation in particular. A number of general conclusions can in fact be drawn from what has been covered in this chapter:

(1) That visual performance improves sharply with increase in lighting level (L_{av}), up to an L_{av} of at least 1 cd/m² (assuming moderate luminance uniformity and glare restriction, viz. U_0 and TI values of around 0.4 and 10 per cent respectively).

(2) That the improvement in visual performance with increase in L_{av} starts levelling out between 2 cd/m² and 10 cd/m² (assuming moderate uniformity and glare restriction).

(3) That at all lighting levels there is a significant worsening of visual performance if either the uniformity (U_0) or the glare restriction (the threshold increment TI) of the installation is degraded.

(4) That there exists a certain interrelation between level (L_{av}), uniformity (U_0) and glare (TI) such that a fall-off in some aspect of performance caused by a degradation of U_0 or TI can be made good by an increase in L_{av} or, conversely, when caused by a decrease in L_{av}, can be more or less rectified by an improvement in U_0 or TI. For example, a drop in

detection probability or revealing power caused by a drop in U_0 from 0.4 to 0.2 can be made good by increasing L_{av} by a factor of about 3.5 to 4, while the factor increase in L_{av} needed to compensate for the drop in visibility index that takes place with this change in uniformity is 2.4.

Similarly, where TI is increased from 7 to 30 per cent, both revealing power and visibility index can be maintained at a constant level by increasing L_{av} by a factor of about 2.5.

(5) That the longitudinal uniformity (U_l), known to influence visual comfort, also influences the reaction performance of motorists.

Chapter 3

Visual Comfort

In order to gain an insight into the way in which the various lighting parameters affect the visual comfort aspect of visual reliability and thereby visual reliability as a whole, the subjective appraisal technique is often used. In this, the members of a relatively large group of observers are each asked to state their opinion on a certain aspect of the lighting known to influence visual comfort, viz. level (L_{av}), uniformity (U_l) and glare (G). The nine-point scale of assessment often used for this purpose is shown in table 3.1 (De Boer, *et al.*, 1959). Details of such subjective appraisal studies carried out in the field and using models are presented in this chapter.

Table 3.1 The nine-point scale used when making subjective appraisals (De Boer, et al., 1959)

Appraisal	Level and Uniformity	Glare
1	Bad	Unbearable
3	Inadequate	Disturbing
5	Fair	Just admissible
7	Good	Satisfactory
9	Excellent	Unnoticeable

Note: Even numbers correspond to appraisals midway between those given above.

3.1 Level

The first subjective appraisal carried out in the field on existing road lighting installations in which the nine-point scale above was used, was carried out in the Netherlands (De Boer, *et al.*, 1959). It revealed that the lighting level was judged to be 'good' (i.e. 7 on the scale of assessment given above) if the road-surface luminance was 1.5 cd/m². As 28 of the 70 installations judged had the character of a secondary road, it is likely that for main roads a value higher than 1.5 cd/m² would be needed for them to be given the same appraisal. Ten years later, in 1969, these assessments were repeated by experienced road lighting engineers (mainly members of the Dutch committee for public lighting) on 49 different road lighting installations, again in the Netherlands (De Grijs, 1972). The appraisals were done by the driver and co-driver of a car travelling at a moderate speed. In order to be able to subsequently distinguish between the results obtained for secondary roads and those applicable to main roads the observers were asked, in part of the tests, to appraise each installation according to the road class concerned.

The results of these 1969 assessments are given in figure 3.1 along with the results of appraisals carried out in 1971 on secondary roads (De Boer, 1972) by substantially the same observers who did the 1969 appraisals. Each point in the figure represents the average appraisal of the total group of observers (14 in 1969 and 16 in 1971). For secondary roads the appraisal 'good' is obtained at an average road-surface luminance of 1.2 cd/m^2, whereas for main roads a luminance of nearly 3 cd/m^2 is required for the same appraisal. The spread of the points is due in part to the influence that the uniformity of the luminance pattern has on the appraisal of the average luminance level.

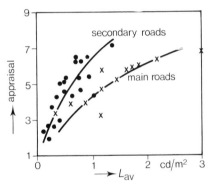

Figure 3.1 Appraisal of lighting level as a function of the average road-surface luminance (L_{av}) for main and secondary roads in the Netherlands. (De Grijs and De Boer)

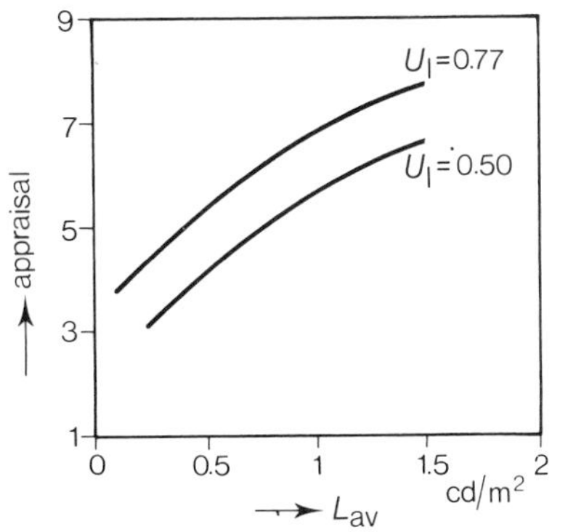

Figure 3.2 Appraisal of lighting level as a function of the average road-surface luminance (L_{av}) for two values of the longitudinal uniformity (U_l). (Economopoulos)

Economopoulos (1978) asked eight observers to appraise the lighting level of two different installations in Greece having longitudinal uniformities of 0.77 and 0.50 respectively. For each installation results for three levels of the road-surface luminance were obtained by repeating the tests with the observers wearing glasses having different neutral transmission factors.

The best-fit curves for the two sets of results are given in figure 3.2. It can be seen that the appraisals for the installation having a longitudinal uniformity of 0.50 are always 1 unit lower than those for the other installation with the better uniformity.

The average appraisals obtained by Economopoulos lie closer to those shown in figure 3.1 for the secondary roads than to those for the main roads, although the roads appraised had in fact more in common with main roads than with secondary roads. This difference between the Dutch and Greek appraisals may be due to a difference in the way the character of the various roads is defined and, more than likely, to a difference between the two groups of observers.

3.2 Uniformity

The subjective appraisals of level conducted by De Grijs and De Boer referred to above were accompanied by subjective appraisals of uniformity, the same roads being used for both sets of appraisals (De Grijs, 1972 and De Boer, 1972).

De Grijs found no correlation between the subjective appraisal of uniformity and the two luminance ratios L_{min}/L_{av} – the overall uniformity ratio important for visual performance – and L_{min}/L_{max}. But subsequent analysis by both De Boer and De Grijs revealed the existence of a correlation between the uniformity appraisal and the ratio L_{min}/L_{max}, provided this was measured along a line parallel to the axis of the road – the longitudinal uniformity (U_l) defined in Chapter 1.

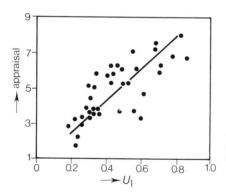

Figure 3.3 Appraisal of uniformity as a function of the longitudinal uniformity (U_l) for both main and secondary roads.

Figure 3.3 gives the results of the above appraisals, which were carried out on two-lane single-carriageway roads, and shows the correlation referred to, the appraisal 'good' corresponding to a longitudinal uniformity (U_l) of about 0.7. The spread in the appraisals apparent from this figure must be partly due to the grouping of the results for the secondary and main roads, but it is also more than likely that differences in luminance level and luminaire spacing are contributory factors. These influences, and the resulting inter-relationships between U_l and L_{av} on the one hand and between U_l and spacing on the other are discussed below.

3.2.1 The influence of L_{av}

De Boer and Knudsen (1963) appraised uniformity on the basis of the relative maximum luminance slope (S_{max}), as defined in Sec. 1.3.2. They conducted their appraisals using a semi-dynamic road lighting simulator, scale 1:50. The simulator, which was also used for the glare appraisals described in Sec. 1.3.3, can be seen in figure 1.6.

Figure 3.4 shows the relation between the uniformity appraisal and the S_{max} value for four values of the average road surface luminance (L_{av}) as obtained using the simulator. The influence of L_{av} on the uniformity appraisal is quite apparent: doubling the former increases the appraisal for a given value of S_{max} by about 1.1 units on the scale. Figure 3.4 shows also that an increase in S_{max}, i.e. an increase in luminance gradient, has a negative effect on the appraisal of uniformity.

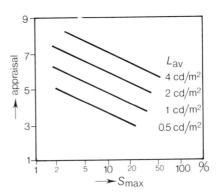

Figure 3.4 Appraisal of uniformity as a function of the relative maximum luminance slope (S_{max}) for four values of the average road-surface luminance (L_{av}). S_{max} is defined as the maximum luminance variation found over any 3 m along or 1 m across the road, expressed as a percentage of the average road-surface luminance.

Economopoulos (1978), from uniformity appraisals conducted in the field, found that the appraisal was not quite so strongly influenced by the level; namely, that doubling L_{av} increased the uniformity appraisal by between 0.5 and 0.8 units.

Returning to figure 3.3, it can be seen that doubling the average road-surface luminance to give uniformity appraisals of +0.5, +0.8 and +1.1 is equivalent to increasing the longitudinal uniformity by 0.06, 0.09 and 0.12 respectively.

3.2.2 The Influence of Spacing

Walthert (1973 and 1975) used a semi-dynamic road lighting simulator similar to that used by De Boer and Knudsen to investigate the influence of the luminaire spacing on the appraisal of uniformity. As the spacing was altered the observers were asked to adjust the longitudinal uniformity (using a mechanical device) to keep it just acceptable. This procedure was repeated for four values of the average road surface luminance.

The results are shown in figure 3.5. It can be seen that as the spacing is increased (the average road surface luminance being kept constant), the uniformity can be decreased without it being appraised any lower. This finding can undoubtedly be attributed to the negative influence that an increase in the luminance gradient is known to have on the appraisal of uniformity, as was illustrated in figure 3.4.

Since Walthert's observers did not appraise the uniformity on a nine-point scale but were asked to make it just acceptable, it is likely that they were in fact setting the luminance pattern so as to obtain a good installation, that is to say one that would be appraised as 7 on the scale of table 3.1. Assuming this to be the case, the value of $U_l = 0.7$ given by figure 3.5 at $L_{av} = 2$ cd/m^2 and $s = 40$ m is in close agreement with the results of De Boer and De Grijs (figure 3.3).

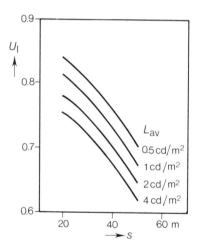

Figure 3.5 Just acceptable longitudinal unifor- mity (U_l) as a function of luminaire spacing (s) for four values of the average road-surface luminance. Driving speed: 50 km/h. (Walthert)

The influence of the luminance level on the assessment of uniformity is less than that found in Sec. 3.2.1. According to figure 3.5, doubling the level decreases the value required for U_l by about 0.03 whereas Economopoulos found the corresponding decrease in U_l to be between 0.06 and 0.09 and De Boer and Knudsen suggest that U_l can be reduced by as much as 0.12.

3.3 Glare

The method used to calculate the glare control mark (G) explained in Sec. 1.3.3 is based on observations carried out on simulated installations created using laboratory models and a test road in an open-air lighting laboratory.

The validity of the method has, of course, been checked by extensive field tests, the results of which are presented in figure 3.6 (De Grijs, 1972). Each point in the figure represents the average of the individual subjective glare appraisals of the observer group plotted against the calculated glare control mark (G) for the installation in question. The observers and the installations are those used for the level and uniformity appraisals described earlier (figures 3.1 and 3.3).

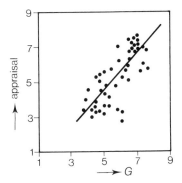 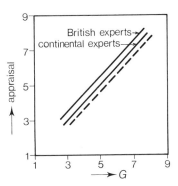

Figure 3.6 Correlation between glare appraisal and calculated glare control mark (G) derived from a number of different lighting installations. (De Grijs)

Figure 3.7 Best-fit lines showing the correlation between glare appraisal and calculated glare control mark (G) for 26 important traffic routes in Great Britain. (Cornwell) The broken line is the correlation found by De Grijs, figure 3.6.

The correlation between glare appraisal and calculated G value is quite clear. As to the spread in the results, it is arguable whether this could be due in part to differences in luminaire arrangement influencing the appraisal. De Grijs himself found no evidence of any correlation between appraisal and luminaire arrangement, but Economopoulos (1978) found the appraisal for different arrangements (opposite, single-sided and staggered) to differ by slightly more than 0.5 on the scale of assessment.

The influence that the 'past history' of the observers (expectancy) has on the appraisal of glare is illustrated in figure 3.7. This gives the best-fit curves for the glare appraisals obtained from two groups of 11 expert observers, one British and the other from the continent of Europe, who examined 26 important traffic routes in Great Britain (Cornwell, 1971).

The observers from the continent, where the glare control imposed on the luminaires used is usually tighter than in Great Britain, appraise the same installations slightly but nevertheless significantly lower than do the British observers. Incidentally, it is interesting to note that the trend of the curves is the same as that of the best-fit curve given in figure 3.6.

3.4 Summing Up

The individual relations between the assessment of certain features of road lighting installations and various lighting parameters have been evaluated. It has been shown that relatively good correlations exist between
(1) The appraisal for level and the measured value of the average road-surface luminance (L_{av})
(2) The appraisal for uniformity and the value of the measured longitudinal uniformity ratio (U_l)
(3) The appraisal for glare and the calculated value of the glare control mark (G).

The appraisal for level appears to be particularly dependent upon the type of road considered, the appraisal differing for main and secondary roads. An appraisal 'good' is obtained for secondary roads at an average road-surface luminance of slightly greater than 1 cd/m², whereas for main roads an L_{av} of more than 2 cd/m² is required for the same appraisal.

The appraisal 'good' for uniformity is obtained at a measured longitudinal uniformity ratio (U_l) of around 0.7. It has been shown that higher U_l values are required for spacings smaller than about 40 m and that lower values can be tolerated for wider spacings. It is important here not to confuse longitudinal uniformity (U_l), which is the lighting parameter associated with visual comfort, and overall uniformity (U_o), the parameter dealt with in the previous chapter when discussing visual performance.

The glare control mark (G) has been shown to give a good indication of the degree of comfort corresponding to a given degree of glare control, the numerical value of G agreeing closely with the corresponding index of appraisal (e.g. $G = 7$ corresponds to the appraisal 'satisfactory'). Again, G, which describes the visual comfort aspect of glare, should not be confused with TI the threshold increment describing the effect glare has on visual performance.

Finally, it has been shown that each of the three correlations mentioned above is influenced to some extent by a second lighting parameter not normally associated with the quality aspect under investigation. Level, for example, is appraised 1 unit lower on the scale of assessment if the longitudinal uniformity is decreased from 0.77 to 0.50. Conversely, the appraisal for uniformity increases by 0.5 to 1.1 units on the nine-point scale when the level is doubled. Glare is likewise influenced by the luminance level, but in this case the effect of L_{av} does not have to be allowed for separately, as it is built-into the formula for the glare control mark.

Chapter 4

Spectra and Visual Reliability

The spectral composition of a light source is important on a number of counts. In the first place, it determines the colour appearance of the source and the way that source will render the colour of objects it illuminates. These source characteristics are, however, only of limited importance for the majority of road lighting applications. Colour appearance does sometimes have a part to play in helping to improve the visual guidance given by a lighting installation (Chapter 15), just as colour rendering becomes a design consideration in connection with public amenity lighting (Chapter 16).

But the spectral composition of the light emitted by a source also has a noticeable influence on the criteria used in road lighting in general to assess the visual performance and visual comfort of road users. This is principally because under monochromatic radiation the eye is able to come to a sharper focus than is possible with multi-line or continuous spectra.

4.1 Visual Performance

Many studies (amongst others those of Jainski, 1960b and Buck, *et al.*, 1975) have shown that contrast seeing is scarcely, if at all, influenced by the spectral composition of the light. Those influences that are found, however, are usually in favour of the spectra of low-pressure and high-pressure sodium lamps rather than that of the high-pressure mercury lamp. Buck *et al.*, for example, used a visibility meter to measure the visibility level (Sec. 2.4.1) of

Table 4.1 *Visibility levels measured under low-pressure sodium (LPS), high-pressure sodium (HPS) and high-pressure mercury (HPM) lighting at* $L_{av} =$ *2 cd/m^2. (derived from Buck, et al.)*

Source	Manikin	Dog	Car exhaust pipe
LPS	26	37	3.9
HPS	32	32	4.2
HPM	19	19	3.2

various objects lighted in turn by low-pressure sodium (*LPS*), high-pressure sodium (*HPS*) and high-pressure mercury (*HPM*) lamps, the road surface luminance being held constant at 2 cd/m². Table 4.1 gives the results of these measurements, the effect of the observers' own headlights having been taken into account.

But the overall visual performance of a motorist is also influenced by his visual acuity, speed of perception, and the time needed to recover the latter after exposure to severe glare. Each of these criteria exhibits a pronounced dependence on the spectral composition of the light.

4.1.1 Visual Acuity

The ability to easily detect the presence of an object does not necessarily guarantee that the object is just as easily identifiable. But an object, once detected, cannot be safely dismissed as being unimportant until it has been properly identified, and this entails the perception of small details of the object. The visual acuity of the road user is therefore important to road safety and in maintaining a continuous traffic flow.

From many sources (for example: Luckiesh and Moss, 1933; Jainski, 1960a; De Boer, *et al.*, 1967; Campbell and Gubisch, 1967) it is known that the visual acuity under low-pressure sodium lighting is significantly better than with other, non-monochromatic light sources.

It has been shown (De Boer, *et al.*, 1967) that, at the lighting levels usual for road lighting, the background luminance under high-pressure mercury lighting must be approximately 1.5 times that with low-pressure sodium in order for it to give the same visual acuity.

This has been confirmed in a more recent study (De Boer, 1974) in which the visual acuities under three different sources were compared: low-pressure sodium, high-pressure sodium and high-pressure mercury. In these laboratory tests, Landolt rings were used as the task object. Twenty-six observers viewed each ring as it was displayed for a period of 0.8 s in the centre of a 3 m × 3 m screen at a distance of 5 m from the viewing position.

The results obtained are shown in figure 4.1 in which the just visible gap size (*D*) of the Landolt ring has been plotted as a function of the homogeneous luminance (L_b) of the screen, with the contrast (*C*) between ring and screen as parameter. (The *D* value is that needed for 80 per cent of the observers to correctly identify the orientation of the ring.) These results clearly demonstrate that for all lighting levels investigated (0.2 cd/m² to 5 cd/m²) and for all contrast values, visual acuity under low-pressure sodium lighting is better than with high-pressure sodium; high-pressure mercury lighting giving the lowest acuity of all.

With the aid of the curves given in figure 4.1 the 'equivalent' background luminances – the background luminances needed with each of the three sources to give equal visual acuity – have been determined.

The ratio of the equivalent background luminance under high-pressure mercury to that under low-pressure sodium (written $R_{eq}HPM/LPS$ for short) and the ratio of the equivalent background luminance under high-pressure sodium to that under low-pressure sodium ($R_{eq}HPS/LPS$) have then been

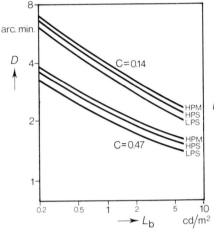

Figure 4.1 Gap size (D) of Landolt ring needed to give 80 per cent correct identification, as a function of background luminance (L_b) for two values of ring contrast.

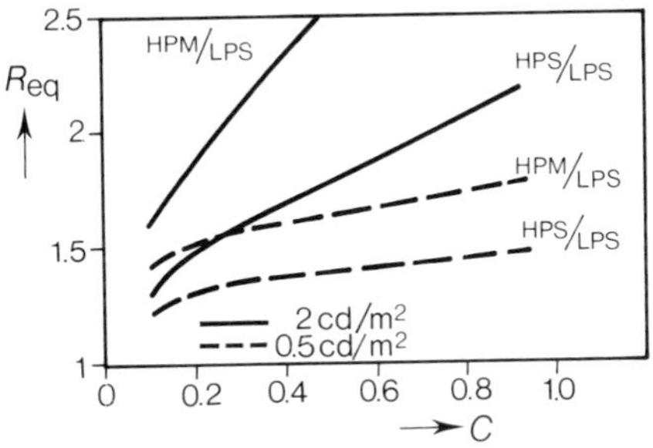

Figure 4.2 Ratio (R_{eq}) of equivalent background luminances for high-pressure mercury and high-pressure sodium lamps, relative to that for low-pressure sodium (0.5 cd/m² and 2 cd/m²), for the same degree of visual acuity, as a function of task contrast (C).

plotted in figure 4.2 as a function of the contrast (C) for two values of the background luminance as given by low-pressure sodium lighting. For the lower of the two background luminances and at low visual task contrasts (the conditions prevailing in the earlier investigation under field conditions) the value of 1.5 for the ratio $R_{eq}HPM/LPS$ is confirmed. But these later investigations go a step further and show that the ratio $R_{eq}HPM/LPS$ is somewhat greater than 1.5 for a road surface luminance of 2 cd/m². Meanwhile, the ratio $R_{eq}HPS/LPS$ can be seen to lie about half-way between unity and $R_{eq}HPM/LPS$ for either of the two road-surface luminances and for any given contrast.

4.1.2 Speed of Perception

Whether or not an object can be seen depends to a large extent also on the time available for its detection – the so-called exposure time. There is a threshold exposure time corresponding to any given combination of object size, contrast and background luminance. The reciprocal of this time is called the speed of perception.

Figure 4.3 shows the speed of perception (S_p) as determined by Weigel (1935) for moving objects (Landolt rings with a gap of 4 minutes forming a contrast of 33 per cent with the background) under low-pressure sodium, incandescent and high-pressure mercury lighting. The background luminance with high-pressure mercury lighting must, at low levels, be approximately 1.5 times higher than that under low-pressure sodium lighting in order for the speed of perception to remain the same. This factor is equal to that required for equal visual acuity (Sec. 4.1.1). Luckiesh and Moss (1936) obtained similar results for speed of perception.

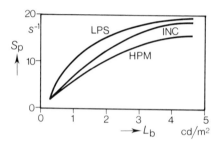

Figure 4.3 Speed of perception (S_p) for a moving object at 33 per cent contrast, as a function of background luminance (L_b), for three types of light source. (Weigel)

4.1.3 Recovery Time

Under practical road lighting conditions it regularly occurs that a driver is briefly exposed to a relatively high level of glare, perhaps due to the fact that one of the lamps or luminaires of the installation is out of its normal position. The time taken to return to an adequate level of visual performance after exposure to glare (often several seconds) is called the recovery time.

Significant differences in recovery time are found for the various light sources employed in road lighting. De Boer (1960), in a laboratory investigation employing six young observers, determined the recovery time on the basis of visual acuity for three types of light source. The test involved viewing Landolt rings at a background luminance of 0.5 cd/m², a glare source being momentarily introduced at an angle of 4° to the line of sight producing an illuminance of 8 lux on the eye of the observer (equivalent veiling luminance: 5 cd/m²). The recovery time was the time needed for an observer to correctly identify the orientation of the ring from the moment the glare source was removed. Table 4.2 gives the average recovery times for the six observers, from which it can be seen that the time under low-pressure sodium lighting is some 20 per cent less than with high-pressure mercury.

Jainski (1962), who measured visual performance on the basis of threshold contrast, found rather greater differences in recovery times. At a background luminance of 0.5 cd/m² and an equivalent veiling luminance of 5.9 cd/m² the recovery time under low-pressure sodium lighting was found to be 40 per cent less than with high-pressure mercury lighting.

Table 4.2 Average recovery time after exposure to glare (De Boer)

Source	Average recovery time (s)
Incandescent lamp (white)	5.7
High-pressure mercury	4.2
Low-pressure sodium	3.4

It should be added that the age of the observer has a definite influence on his recovery time. By way of example, the recovery time after glare for a group of eight persons aged between 50 and 62 years was on the average twice that found for a group of eight persons aged between 18 and 25 years.

4.2 Visual Comfort

The spectral composition of the light in a numer of existing road lighting installations has also been found to have a marked influence on the visual comfort of road users.

4.2.1 Level

As has been mentioned already in Chapter 3, there is a correlation between the average road surface luminance and the subjective appraisal of lighting level. The results of the appraisal test carried out by De Boer *et al.*, which are mentioned in that chapter, are given again in figure 4.4, but this time subdivided according to the type of light source used, viz. low-pressure sodium (*LPS*) and high-pressure mercury (*HPM*).

According to this figure, at the luminance levels common in road lighting (viz. above 0.5 cd/m²) the road surface luminance should be higher under high-pressure mercury lighting than under low-pressure sodium for it to be given the same appraisal: for example, to be appraised as 'good' the luminance should be 1.6 times higher under mercury lighting.

De Boer (1974), in a more recent study, also took into account the effect of high-pressure sodium lighting on the appraisal of lighting levels. In this study, which was carried out using a simulated road lighting installation, observers were confronted with a road lit by high-pressure mercury lamps and having in turn four values of average road-surface luminance: 0.5 cd/m², 1.0 cd/m², 2.0

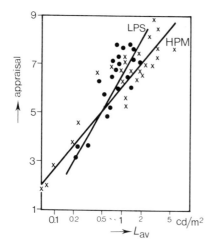

Figure 4.4 Appraisal of lighting level as a function of average road-surface luminance (L_{av}) subdivided according to type of light source employed.

cd/m² and 4.0 cd/m². At each luminance the observers were asked to adjust the level of the low-pressure sodium lighting and then the high-pressure sodium lighting, in place of the original high-pressure mercury, so as to keep the brightness of the road surface constant throughout. In line with the results of figure 4.4, only slight differences between adjusted luminances were found at the two lower values of road surface luminance, but at 2 cd/m² and at 4 cd/m² these differences became significant, table 4.3.

The difference is not so great as was found in the previous test referred to above (1.09 as against 1.60 for the ratio $L_{av}(HPM)/L_{av}(LPS)$, but this may be attributable to the fact that the criteria used to assess the level were different, viz. visual comfort in the first case, and surface brightness in the second. This later study shows also that an even greater reduction in road surface luminance is acceptable when using low-pressure sodium lighting in place of high-pressure sodium.

4.2.2 Uniformity

Field tests on completed road lighting installations have failed to reveal any link between the type of light source used and the appraisal obtained for

Table 4.3 Ratios of those average road-surface luminances giving the same impressions of surface brightness (De Boer)

Surface luminance (HPM)	$L_{av}(HPM)/L_{av}(LPS)$	$L_{av}(HPS)/L_{av}(LPS)$
2 cd/m²	1.09	1.16
4 cd/m²	1.15	1.20

uniformity. An investigation carried out with the dynamic road lighting model shown earlier in figure 1.6 also failed to point to any significant connection between these two factors.

4.2.3 Glare

Disability glare is in no way influenced by the type of light source employed (Jainski, 1962), but the same cannot be said for discomfort glare. De Boer, *et al.* (1955) determined the just admissible luminaire luminance for both low-pressure sodium and high-pressure mercury sources, L_{LPS} and L_{HPM} respectively. The average value of the ratio L_{LPS}/L_{HPM}, as determined by a group of 50 observers in a laboratory set-up, was found to be as high as 1.45 at an average road surface luminance of 1 cd/m^2. A year later, Ferguson and Stevens (1956) arrived at a figure of 1.7 for this ratio in similar tests conducted in the field.

Figure 4.5 Photograph of lighted road referred to in the text which was employed when making assessments of discomfort glare from various light sources.

Further laboratory investigations have since been carried out incorporating the results for high-pressure sodium luminaires (De Boer, 1974). Twenty observers were seated in turn at a distance of 1.5 m from a photograph of a lighted road (figure 4.5) which had been enlarged to such a size that it truly represented the view of the road as would normally be seen by a driver. Holes were made in the photograph at the positions occupied by the luminaires, and in each hole an image was projected of either high-pressure mercury, high-pressure sodium or low-pressure sodium luminaires. This was done in such a way that the observer saw each hole completely and homogeneously filled with the luminance of the light source chosen. A system of movable mirrors and shutters was employed to produce rapid changes in the lighting situation – changeover time about 1/10 s. Each observer was asked to adjust the luminance of the 'luminaires' to give first a 'just admissible' and then a 'satisfactory' degree of glare for each situation.

The results are given in table 4.4 from which it can be seen that glare would again appear to be less of a problem with low-pressure sodium than with high-pressure mercury luminaires, the figures for the ratio L_{LPS}/L_{HPM} being close to the 1.45 and 1.7 mentioned above. Given that the luminaire luminance may be higher with low-pressure sodium than with high-pressure mercury (which means that for equal luminous areas the intensity may also be greater by the same amount) it is reasonable that a correction should be applied to the glare control mark as calculated using the general formula given earlier (Sec. 1.2.3). Calculation in fact indicates that a correction to the calculated G value of between $+0.5$ and $+0.8$ may be applied (L_{LPS}/L_{HPM} equals 1.45 and 1.7 respectively) when using low-pressure sodium luminaires.

Table 4.4 Ratios of those luminaire luminances giving the same degree of glare (De Boer)

Degree of glare	L_{av} (cd/m^2)	L_{LPS}/L_{HPM}	L_{LPS}/L_{HPS}
'Satisfactory'	1	2.1	1.1
	2	1.5	1.1
'Just admissible'	1	1.8	1.2
	2	1.6	1.1

From table 4.4 it can be seen that the differences between low and high-pressure sodium with regard to glare are relatively small. This being the case, high-pressure sodium light must be less glare inducive than that from a high-pressure mercury source, although the advantage is slightly smaller than that shown by low-pressure sodium over mercury. According to the figures given in this table, the correction for G when using high-pressure sodium should be 0.2 smaller than the correction for low-pressure sodium (viz. between $+0.3$ and $+0.6$).

The CIE (1976b) is rather conservative in its adoption of correction factors for the various light sources. It specifies that for low-pressure sodium lamps the calculated glare control mark may be increased by 0.4 (i.e. colour correction factor, C, in glare formula equals 0.4), while for all other light sources the correction is zero (i.e. $C = 0$).

4.3 Summing Up

It has been shown that the spectral compositions exhibited by the light sources commonly employed in road lighting can have an important influence on the performance criteria used in assessing visual performance and on the lighting parameters employed to assess visual comfort.

Visual performance

With regard to visual performance, there is a clear change in each of the three criteria: visual acuity, speed of perception and glare recovery time when going from high-pressure mercury lighting to that provided by either high-pressure sodium or low-pressure sodium lamps. Put another. way, the amount of light that must be provided by each of these sources to obtain the same level of performance is different, viz.

Visual acuity. For this to remain constant, the background luminance provided by high-pressure mercury lighting must be at least 50 per cent higher than that given by low-pressure sodium. Similarly, the background luminance under high-pressure sodium lighting needs to be at least 25 per cent higher than that provided by low-pressure sodium.

Speed of perception. For equal speed of perception, the background luminance under high-pressure mercury lighting must again be at least 50 per cent higher than that provided by low-pressure sodium.

Glare recovery time. The time needed to return to a given level of visual performance after exposure to glare is at least 20 per cent shorter for low-pressure sodium glare sources than for high-pressure mercury.

Visual comfort

The visual comfort given by various light sources was assessed in terms of the road surface luminance and luminaire brightness, the assessments being set against the measured values of the two lighting parameters level (L_{av}) and glare (G).

Level. For the assessment of this to be the same under both high-pressure mercury and low-pressure sodium lighting the average road surface luminance given by the former must be some 10 to 60 per cent higher than that given by the latter. Similarly, for equal assessments of level under high-pressure sodium and low-pressure sodium, the luminance given by the former must be 15 to 20 per cent higher.

Glare. For this to be appraised as 'satisfactory', the luminance of low-pressure sodium luminaires may be up to 50 per cent higher than that of high-pressure mercury luminaires, but only 10 per cent higher than that of high-pressure sodium luminaires.

Chapter 5

Tunnel Lighting

When discussing the basic considerations governing the lighting of road tunnels a distinction has to be made between those requirements that are applicable to the night-time lighting of all tunnels, regardless of length, and those special, day-time requirements that relate only to tunnels above a certain critical length.

At night, a tunnel may be lighted in much the same way as would a similar stretch of open road, and as far as the visual performance and visual comfort aspects of the lighting are concerned, most of what has been said in the previous four chapters concerning the requirements of normal road lighting is again applicable in the case of tunnels. Suffice it to say here that the basic difference between the two sorts of installation is that tunnel lighting calls for somewhat higher lighting levels than are normally recommended for normal road lighting. The reason for this is that driving through a tunnel is normally considered to be more hazardous and demanding than driving along an open road – there is less room to manoeuvre in the event of an accident and a greater degree of disturbance from noise sources due to the tunnel acoustics.

Compared with the night-time requirements of tunnel lighting, the requirements that must be satisfied in critical-length tunnels during the hours of daylight are considerably more demanding.

The problem by day is principally one of adaptation, and is connected with the fact that when the human eye is adapted to daytime brightnesses it is quite unable to function properly when suddenly confronted with relatively low luminances in a tunnel entrance, luminances that would be quite adequate for a dark-adapted eye at night. For this reason, the lighting in the first part of a tunnel that qualifies for extra lighting must be related to the adaptation state of the eye on entry, and so arranged that the subsequent transition from the highest to the lowest luminance levels, which is done in the interests of economy, is made gradually, in order to give the eyes time to adapt.

Before going on to examine some of the most important findings to emerge from investigations carried out into this and other aspects of tunnel lighting, it is necessary to explain what is meant by the term 'critical length'. Broadly speaking, a tunnel is said to be shorter than the critical length and generally not in need of lighting if the day-lit exit occupies a quite large part of the field of view when seen from a distance equal to the driver stopping distance before the tunnel entrance. Conversely, it can be considered to be longer than the critical length, and therefore in need of extra daytime lighting if, when viewed from the same distance, it appears as a dark frame round the bright centre of the exit (figure 5.1).

Figure 5.1 A tunnel shorter than the critical length and that from this direction of approach would appear not to require daytime lighting since the bright exit is clearly visible and obstacles would be silhouetted against this in time for avoiding action to be taken (but see figure 5.3).

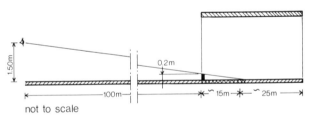

Figure 5.2 Sketch illustrating how the critical tunnel length of 40 metres is derived (see text).

With certain exceptions (see below), the critical length is about 40 m. This length is arrived at as follows (figure 5.2). Daylight entering a tunnel via the exit will lighten the road surface, sufficient to make an object on it visible, over a distance of about 25 m (Schreuder, 1964 and Narisada and Yoshikawa, 1974). This lightened road surface will form a suitable background against which the top of a 'standard' 20 cm high object at the tunnel entrance will be clearly silhouetted if, when viewed from a point 100 m in front of the tunnel and 1.5 m above the road surface, the distance between the object and the beginning of the lightened road surface is not greater than 15 m.

Figure 5.3 The tunnel of figure 5.1 but seen from the other approach. The tunnel must, in fact, be provided with daytime lighting because from this approach the bright exit is not visible until just before entry.

This 40-m-rule applies only to straight level tunnels not carrying a heavy traffic loading. If there is a curve in the tunnel or in the tunnel approach road (figure 5.3), if this dips, or if sight of the exit is frequently lost due to the presence of intervening vehicles, then special tunnel lighting should be provided.

5.1 Threshold Zone

The eyes of a driver approaching a tunnel entrance are adapted to the high level of day-time luminance. Consequently, if the luminance level inside the tunnel is much lower than that outside, no details of its interior or any objects there will be visible – in extreme cases, the tunnel entrance may even appear from outside as a 'black hole'.

The first of the four zones of a long tunnel (figure 5.4) requiring special lighting during the hours of daylight in order to maintain the visual reliability of an approaching driver at an acceptable level, is the threshold zone. The walls and road surface in this zone form the background against which obstacles in his path are seen.

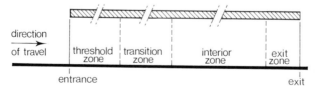

Figure 5.4 Tunnel zones distinguishable for lighting purposes.

62

5.1.1 Luminance Level

The luminance level needed in the threshold zone to maintain the visual performance of the road user at a safe level is dependent upon the severity of the visual task likely to be encountered and on the luminances prevailing outside the tunnel entrance.

In most investigations connected with tunnel lighting the visual task has been standardised as consisting of an object measuring 20 cm × 20 cm which has to be visible from a distance of 100 m (i.e. angle subtended at eye of observer equals 7 minutes of arc). This task specification is in line with the considerations outlined earlier in Chapter 2 – Visual Performance.

The influence that the outside adaptation luminance has on the luminance needed in the threshold zone in order to see the task object has been investigated by considering the case of (a) a uniform adaptation luminance and a stationary observer, and (b) a non-uniform adaptation luminance and an observer on the move. (Farther on in this Section it will be shown that the results for a stationary observer and a uniform luminance can in fact be used as a basis for determining the required threshold zone luminance for a moving observer and a non-uniform luminance of the tunnel surround, which are the conditions actually occurring in practice.)

Stationary observer and uniform luminance

Schreuder (1964 and 1967) used a static laboratory set-up to determine the luminance in the threshold zone, as a function of the uniform luminance surrounding the tunnel entrance. The object, subtending an angle of 7 minutes of arc, was presented to the observers for 0.1 s, a presentation time which has subsequently been shown to be realistic in investigations into the eye movements actually performed by drivers. (Narisada and Yoshikawa, 1974a, using a corneal eye-marker recorder fixed to the head of a driver, found that the shortest period for which the eye is stationary is between 0.1 s and 0.2 s.)

The object was displayed in the centre of a small rectangle located in the centre of a large screen of uniform luminance which extended over the observer's entire field of vision, figure 5.5. The small rectangle (*TH*) of luminance L_{th} represented the threshold zone of a tunnel as seen by a driver some distance from the tunnel entrance and the screen (*A*), of luminance L_a, the adaptation luminance. Prior to the object being presented, it and *TH* were covered by a shutter, such that the whole screen was of luminance L_a. The values of L_a, L_{th} and *C* (object contrast) could be varied independently of one another.

From the results, the value of L_{th} at which the object was just visible, in 75 per cent of the presentations, to an observer fully adapted to L_a, was plotted for various combinations of L_a and *C*, figure 5.6.

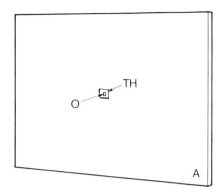

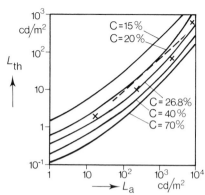

Figure 5.5 Sketch of screen (A) used in Schreuder's static laboratory set-up showing transparent glass slide (TH) with at its centre the object (O).

Figure 5.6 Threshold-zone luminance L_{th} needed to make an object of contrast C just visible, as a function of the uniform adaptation luminance L_a:
solid lines – Schreuder
broken line – Mäder and Fuchs (C=20%)
crosses – Narisada and Yoshimura (C= 27.6%)

Mäder and Fuchs (1966) also conducted investigations along these lines. The object used had a contrast of 20 per cent with its background, but was larger (viz. 80 cm × 40 cm) which gave a subtended viewing angle of 34 × 17 minutes of arc at 80 m. This would automatically have led to lower values for the threshold luminance, but the detection probability called for was increased from 75 to 99 per cent and this had the opposite effect on the luminance. Their results for the luminance range investigated, which can be given by the formula

$$L_{th} = 0.15\, L_a^{0.9}$$

are shown in figure 5.6 by the broken line. As can be seen, the threshold luminance levels are slightly lower than those indicated by Schreuder's curve for a 20 per cent contrast value, especially for the higher outside adaptation luminances.

Narisada and Yoshimura (1974) carried out similar investigations under similar conditions and obtained results that are again in line with those of Schreuder. The results for an object with a contrast of 27.6 per cent have been indicated in figure 5.6 by crosses, and these are very close to those of Schreuder for the object with a contrast of 26.8 per cent.

It is evident from figure 5.6 that object contrast has an important bearing on the lighting level required in the threshold zone. It is necessary, therefore, to examine carefully the lowest value of contrast that could conceivably be expected to occur in actual tunnels.

The luminance contrast of an object is dependent upon the reflectance of the

64

object concerned, and also upon the combination of the reflection properties of the road surface and walls forming the background to the object, the type of lighting arrangement and the luminous intensity distribution of the luminaires employed.

This combination can, as far as the resulting contrast is concerned, be characterised by the ratio of the vertical illuminance on the object to the background luminance, E_v/L_b. It has been shown (Narisada and Yoshikawa, 1975) that for tunnel lighting conditions a value for this ratio of about 7 lux/cd/m² may be regarded as the highest possible and most unfavourable. With counter-cast lighting installations a considerably lower, more favourable value is possible (see also Chapter 17 – Tunnels and Underpasses).

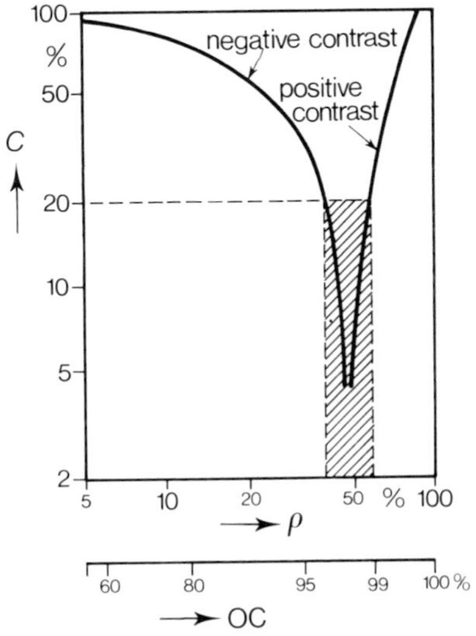

Figure 5.7 Relation between diffuse reflectance ρ of an object and the resulting contrast C for tunnel lighting installations producing an E_v/L_b ratio of 7. The shaded area indicates the reflectance range over which the contrast is less than 20 percent. The lower scale (OC) gives the percentual cumulative occurrence of the reflectances of pedestrians' clothes.

Figure 5.7, which is taken from draft CIE Technical Report: Luminance in the Threshold Zone (CIE, 1979b), gives the variation in luminance contrast over a range of values of diffuse object reflectance ρ for an E_v/L_b ratio of 7 lux/cd/m². The reflectance scale is accompanied by a scale showing the percentual cumulative occurrence of the reflectances of clothes worn by pedestrians (previously used in Chapter 2). From the figure it appears that all contrast values may occur and that there is thus no theoretical method to precisely define the lowest value of contrast occurring in tunnels. It can, however, be seen from the figure that a contrast of 20 per cent, which is often taken as the critical value for tunnel lighting, is a reasonably representative choice for practical purposes. (The shaded area in the figure representing the reflectance range leading to contrasts lower than 20 per cent indicates that only about 3 per cent of pedestrians' clothes produce contrast values lower than 20 per cent.)

65

Based on what has been said above, and using Schreuder's curve for $C = 20$ per cent (figure 5.6), it is possible to determine the luminance needed in the threshold zone as a function of the uniform outside adaptation luminance. For convenience, the top part of this curve covering the adaptation luminances most likely to occur in daytime can be approximated by the straight line of slope 0.10 indicated in figure 5.8. This figure shows that for outside adaptation luminances greater than about 100 cd/m², the lighting level (L_{th}) in the threshold zone should be equal to or greater than 0.10 times the adaptation luminance (L_a). If, instead of 20 per cent, one assumes a critical contrast of 25 per cent, the corresponding requirement for L_{th} (from figure 5.6) becomes: $L_{th} \geqslant 0.07 L_a$.

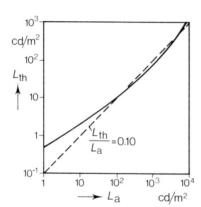

Figure 5.8 The luminance L_{th} needed in the threshold zone, as a function of the outside adaptation luminance. L_a, based on a critical object contrast of 20 per cent.

Moving observer and non-uniform luminance
The results given so far are in principle only valid for a uniform adaptation luminance and a stationary observer. But in practice, of course, the luminance of the surroundings of a tunnel entrance are by no means uniform and the situation as viewed by an approaching driver is far from static.

Moving observer. During his approach to a tunnel the relatively dark tunnel entrance and its surround occupy an increasingly larger part of a driver's total field of view. As a consequence of this, the state of adaptation of the driver's eyes is continuously changing. However, at any given moment during the approach his state of adaptation is determined not only by the luminance distribution in his field of view but also, because of 'time lag', partly by the luminance distribution experienced shortly before.
The present CIE Recommendations for Tunnel Lighting (CIE, 1973b) suggest that for the purposes of arriving at the correct threshold zone luminance the outside adaptation luminance should be determined from the luminance

distributions occurring in a specified field of view at three distances from the tunnel entrance. Further, that these distances should be specified according to the speed limit in force for the tunnel concerned. This approach is further explained in Chapter 9 – Recommendations. The recommendation is, however, on the point of being revised: the CIE is considering recommending in future that the luminance at the beginning of the threshold zone be determined on the basis of the luminance distribution occurring in a driver's field of view at the moment when he has to be able to see a critical object in the tunnel (Van den Bijllaardt, 1980). For an object standing in the mouth of the tunnel this moment is defined by the 'stop decision point', this being the point outside the tunnel at which a driver has to begin braking in order to come to a standstill at the tunnel entrance.

Figure 9.6 in Chapter 9 gives stopping distances for a range of driving speeds (the reaction time having been taken into account) for a wet road surface and various gradients. For example, for a speed of 80 km/h on a level road the safe driver stopping distance is 100 m. Thus, in practice, the adaptation luminance on which the luminance level in the threshold zone should be based can still be determined for a stationary observer, his stop decision point being varied according to the speed limit in force.

Non-uniform luminance distribution. In the context of tunnel lighting, the adaptation luminance for an observer whose view is fixed on a non-uniform luminance distribution is defined as that value of uniform luminance in front of the observer that would result in the same degree of perceptibility as with the non-uniform distribution actually prevailing. (Perceptibility, for this purpose, is defined on the basis of detectable contrasts.) This adaptation luminance is not a physical quantity, but a luminance obtained via physiological equivalence; the adaptation luminance for a non-uniform luminance distribution is therefore more properly called the equivalent adaptation luminance From the definition, it follows that the results of the investigations carried out with the uniform luminance distribution, described above, apply in full to non-uniform luminance distributions if for 'adaptation luminance' is read 'equivalent adaptation luminance'.

The question of how the equivalent adaptation luminance should actually be determined from the luminances in the field of view of a driver at the stop decision point, has been the subject of several investigations.

Schreuder (1964) assumed that the equivalent outside adaptation luminance (especially for road tunnels under rivers, where no special measures are taken to lower the outside luminances) is essentially determined by the luminance of the road surface immediately in front of the tunnel. The luminance of the usual types of asphalt road surface can reach values as high as 8000 cd/m² with a horizontal illuminance on the open road of 100000 lux. An outside adaptation luminance of 8000 cd/m^2 is therefore often used in recommendations on tunnel lighting as a reference value for lighting design purposes. Narisada and Yoshikawa (1974), however, found that drivers concentrate visually for about 80 per cent of the time on the tunnel entrance itself during the last 150 m of their approach. Consequently, the relatively low luminance of the tunnel entrance must have an important influence on the final equivalent adaptation luminance, possibly causing it to be considerably lower than the 8000 cd/m^2 reference value stated above.

But the road surface leading up to the tunnel entrance and the tunnel entrance itself are not the only areas entering the driver's field of view; the sky too can still form a substantial part of this, and its luminance can be as high as 50000 cd/m^2. Unless screened from view by a special canopy or other device (see Chapter 17) such a bright sky will also exert an important influence on the final equivalent adaptation luminance, although this is less likely to be the case for tunnels in mountainous areas and built-up areas, since here the sky will generally occupy only a small part of the field of view.

The CIE (1973b) suggests that these various influences can be allowed for by basing the equivalent outside adaptation luminance on the average of the luminances contained in a conical field of view subtending an angle of $20°$ at the eye and centred on a horizontal viewing direction in the direction of travel. Narisada *et al.* (1979) states, that this CIE approach gives an approximation of the equivalent adaptation. Very recently he concluded that about 1.5 times the average luminance in the $20°$ cone gives the best approximation.

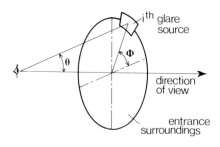

Figure 5.9 The tunnel entrance surround can be considered as consisting of small glare sources, each with an actual measured luminance L. From these luminances the equivalent veiling luminance L_v can be calculated using

$$L_v = k \iint L(\Phi, \Theta) \frac{\sin \Theta \cos \Theta}{\Theta^2} \, d\Phi \, d\Theta$$

where k is the age factor (see Sec. 1.2.3) and L the luminance of a source whose position is defined by the angles Φ and Θ.

It is anticipated that the equivalent outside adaptation luminance can be very accurately determined by way of a technique that is proposed by Adrian (1976) that considers the influence that the tunnel surroundings have on light scatter within the eye. This involves calculating the equivalent veiling luminance from the non-uniform tunnel entrance surroundings, which are treated as consisting of small glare sources each with a luminance equal to the luminance of the corresponding part of the actual surrounding (figure 5.9).

At the time of writing, a generally accepted method for determining the equivalent adaptation luminance based on this approach has not yet been derived, although the preliminary results look promising (see Narisada *et al.*, 1979 and Adrian, 1980).

Luminance decrease in the threshold zone

The threshold zone luminance is at present often kept constant, at the level required at the start of the tunnel, throughout the length of the zone. However, the fact that a driver's eyes usually become gradually adapted to lower luminances between the stop decision point and the relatively dark tunnel entrance due to the increase in the apparent size of the latter, means that this luminance can in fact be allowed to decrease at the same rate as the adaptation luminance.

At the time of writing, however, no accurate information concerning the adaptation state of a driver who is much closer to the tunnel entrance than the stop decision point is available – 'time lag' probably plays an important role here. Further research is therefore needed in this direction before any definite guidance can be given as to what would constitute a safe rate of luminance decrease in the threshold zone. In the meantime, bearing in mind that complete adaptation from high to low lighting levels is a slow process, it is to be recommended that the present practice of maintaining the luminance in this zone at a constant level be continued.

5.1.2 Threshold Zone Length

Threshold lighting serves to produce a background luminance against which an obstacle anywhere on the road surface within the zone will be clearly visible to a driver approaching the tunnel entrance. The lighting should be such that this facility extends over a distance equal to the safe driver stopping distance measured into the tunnel from the entrance. This calls for a threshold zone whose length must be equal to this stopping distance plus an extra distance sufficient to provide a background for an object.

The obstacle chosen for the purposes of determining the extra threshold zone length needed is the familiar plate measuring 20 cm × 20 cm. Looking again at figure 5.2, it can be seen that such an obstacle viewed from a distance of 100 m and from a height of 1.5 m requires 15 m of road surface to form a suitable background. Thus, for a driving speed of 80 km/h corresponding to a safe driver stopping distance of 100 m, the threshold zone must have a length of at least 115 m. For other speeds (see figure 9.6) the length of the threshold zone will vary accordingly.

5.2 Transition Zone

After the relatively high lighting levels required in the threshold zone, the lighting in the tunnel may be gradually reduced to a much lower level. The zone in which this reduction takes place is called the transition zone (see figure 5.4).

5.2.1 The Luminance Reduction Curve

From a point somewhere between the stop decision point and the tunnel entrance the luminance contained within a driver's field of vision starts to decrease rapidly and this causes a change in the adaptation state of the eyes. If this decrease, which is determined principally by the luminances in the transition zone, is too rapid, visibility and visual comfort will deteriorate due to adaptation deficiency.

A principal effect associated with adaptation deficiency is the occurrence of after images. These are weak images of visual scenes that have been witnessed a short time previously. The after image is seen in reversed contrast and colours. The luminance gradient in the transition zone that can be tolerated without it giving rise to the occurrence of after images has been the subject of investigation.

Schreuder (1964) employed a static laboratory set-up in which observers were asked to reduce the luminance of a large screen as rapidly as possible without losing sight of a test object on it and without experiencing after images. The observers were given two minutes in which to allow their eyes to adapt to the initial luminance of $8\,000$ cd/m^2. The resulting luminance reduction curve, admissible in 75 per cent of the observations, is given in figure 5.10.

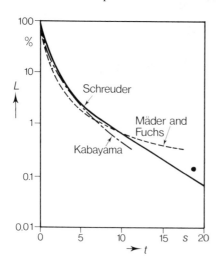

Figure 5.10 Luminance reduction curves showing the maximum permissible rate of decrease in transition-zone luminance L with time t.

70

a

Figure 5.11 Dynamic set-up used by Schreuder to check the validity of his luminance reduction curve: (a) General view showing the seated observer who is moved along rails whilst viewing the interior of the model tunnel through a periscope

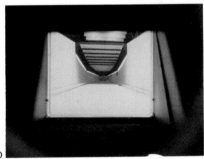

(b) Field of view experienced by observer.   b

Kabayama (1963) and Mäder and Fuchs (1966) conducted similar investigations, but solely on the basis of visibility criteria rather than on both this and the avoidance of after images. Their results are in close agreement with those of Schreuder (figure 5.10).

Schreuder (1964) checked his investigation, described above, using a scale model of a tunnel (described in detail by Balder and Schreuder, 1959) through which the observer actually moved (figure 5.11). The results, part of which are indicated in figure 5.10 by the single point, were in close agreement with those obtained using the former, static set-up.

Schreuder (1964), in a further investigation, found that this test object would remain visible and after images avoided if the luminance were decreased at a rate dependent upon the value of the initial luminance: the lower the initial luminance within the range 10000 cd/m^2 to about 2000 cd/m^2, the quicker could the luminance be reduced, but for initial luminances below about 2000 cd/m^2 the luminance has to be reduced more slowly with decreasing initial luminance. Since, however, little detailed information is available on the dependence of the permissible luminance reduction rate on the initial luminance on the basis of avoidance of both after images and loss of visibility, the luminance reduction curve given in figure 5.10, which is obtained for an initial luminance of 8000 cd/m^2, seems to form a reasonable, general basis for determining the luminance gradient in the transition zone of a tunnel.

5.2.2 Applying the Luminance Reduction Curve

Having seen how the luminance reduction curve given in figure 5.10 has been derived, the next step is to consider how it should be applied in practice for the purposes of determining the lighting needed in the transition zone.

The observers in the experiments from which the luminance reduction curve was derived were exposed to the initial luminance just prior to the luminance being reduced. This being the case, the zero-point of the horizontal time axis of the reduction curve should be made to coincide with the moment just before which the luminance to which the eye is adapted undergoes a sudden decrease. Meanwhile, the 100 per cent point on the vertical axis (representing the initial luminance) should correspond to the value of the luminance determining the adaptation state of the eye at this same moment in time.

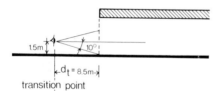

Figure 5.12 Transition point: the point in front of a tunnel entrance from which this just occupies the entire field of view of an approaching driver.

The moment at which these events take place is close to the moment, just before entering the tunnel, when the bright surroundings, *including the road surface in front of the tunnel*, disappear from the view of the approaching driver. The point at which this occurs, viz. the point at which there is a sudden marked transition in the luminance determining the adaptation state of the eyes, can be called the 'transition point', and its distance in front of the tunnel is fixed by the eye height and angle of vision of the driver. Figure 5.12 shows that for the standardised eye height of 1.5 m and an angle of vision of 10°

above and below the horizontal the transition point lies 8.5 m in front of the tunnel entrance. Taking into account variations in the eye height, a fixed distance of 10 m in front of the entrance can safely be assumed for the position of the transition point. However, since it is probable that a driver concentrates his attention most of the time on the road and its surround at a distance equal to his stopping distance ahead of him, it follows that this will be the point at which the luminance reduction should begin to take place. In other words, the zero-point on the reduction curve should coincide in practice not with the transition point itself, but with a point at a distance equal to the safe driver stopping distance beyond it.

The adaptation luminance at the transition point is in fact equal to the equivalent adaptation luminance at this point. As mentioned before, however, there is as yet no accurate information on how the value of the latter should be determined in practice for drivers who are much closer to the tunnel entrance than the stop decision point. Until such information does become available, it is considered safest to use the value of the equivalent adaptation luminance at the stop decision point as the 100 per cent point for the reduction curve.

Figure 5.13 illustrates how the above theory is put into practice. In this

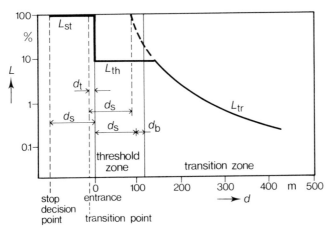

Figure 5.13 Permissible course of the luminance in a tunnel, for a driving speed of 80 km/h, relative to the equivalent adaptation luminance at the stop decision point, where
L_{st} = equivalent adaptation luminance at the stop decision point
L_{th} = threshold-zone luminance
L_{tr} = transition-zone luminance
d_s = safe driver stopping distance (100 m at 80 km/h)
d_t = distance of transition point from tunnel entrance (10 m)
d_b = distance between standard 20 cm square object and end of threshold zone needed to form an object background assuming a standard eye height of 1.5 m and a viewing distance equal to the safe driver stopping distance.

73

example, the threshold zone luminance has been made equal to 10 per cent of the equivalent adaptation luminance at the stop decision point. The time scale of the original reduction curve has been transformed here into a scale of distance travelled at 80 km/h. The luminance reduction for this speed should therefore begin at a point 100 m from the transition point, which is the corresponding safe driver stopping distance. But in practice the luminance in the first part of the transition zone is maintained at the level required at the end of the threshold zone before being gradually reduced as specified by the reduction curve. Schreuder (1964) found that this way of starting the luminance reduction has no influence on the further course of the luminance reduction curve.

In practice, the smooth luminance reduction curve can, in the transition zone, be stepwise approximated. From tests on the admissible luminance jump, Schreuder found that if succeeding steps have transitions of less than 3 : 1 after images will not occur. Needless to say, no step should take the lighting level below that allowed by the curve.

5.3 Interior Zone

In long tunnels, the transition zone is usually followed by a zone in which the luminance level is kept constant. In this, the interior zone, adaptation is not necessarily complete and it is therefore necessary to arrange for a level of luminance which is fairly high compared with the level needed on an open road at night. Even at night, when adaptation is generally more thorough, the lighting level in a tunnel needs to be higher than that on the open road. The reason for this is that in a tunnel there is less space to correct an error in the judgement of lateral distance, less space to avoid obstacles; and if an accident does occur, the consequences are likely to be more severe.

So far as is known by the authors, the lighting level needed in the interior zone has never been the subject of a special investigation, although the levels prevailing in many tunnel interiors have been recorded. Practical experience, however, has shown that during daytime, levels of 5 cd/m² to 20 cd/m² are to be recommended, the lower levels being acceptable for tunnels longer than 1 kilometre, where more time is available for adaptation than in short tunnels. Where large amounts of dust and smoke (exhaust fumes) are to be expected, the lighting level should be high enough to compensate for the loss of visibility caused by this.

5.4 Exit Zone

Since adaptation from low to high levels of luminance takes place practically instantaneously, and because obstacles in the exit zone will stand out clearly

74

against the bright exit, no extra adaptation lighting in the exit zone is called for.

At night, the danger of the black-hole effect occurring at the tunnel exit is present, especially if the open road leading from the tunnel exit is not lighted. In order to avoid this problem, and aid adaptation, transition lighting along this stretch of the road is required. The lighting should cover a stretch of 200 m to 300 m, and should gradually decrease in level, in steps not greater than 3 : 1, to the level provided by the normal road lighting. (For the transition from normal road lighting to an unlit stretch of road, see Chapter 15.)

5.5 Flicker

Light sources mounted in interrupted rows along the ceiling or walls of a tunnel may produce a disturbing flicker in the eyes of a driver passing through it. The flicker is caused both by the light from the sources themselves, which appear and disappear at the edge of the driver's field of view, and by reflections of the sources glimpsed in shiny surfaces, e.g. the bonnet of the driver's own vehicle and the rear of any vehicle he may be following.

The extent to which a driver is troubled by such flicker is largely dependent upon the following factors: the number of flickers occurring per second (flicker frequency); the luminance of the source relative to its background (amplitude); the average luminance in the driver's field of view; the length of the source relative to the spacing; and finally, the total duration of the flicker. Flicker frequency is dependent upon the speed of travel through the tunnel and the spacing of the sources. Jantzen (1960) and Schreuder (1964) both obtained detailed information on the flicker occurring under typical tunnel lighting conditions. With the aid of a model of a tunnel that allowed the motion of an observer to be simulated by moving the sources towards him, they determined, amongst other things, the range of flicker frequencies for which the disturbance occurred.

Jantzen found this range to extend from 3.5 to 15 pulses per second (pps), with a maximum disturbance occurring at 8.5 pps while Schreuder, employing a somewhat lower road surface luminance (8 cd/m^2) and a higher source luminance (9000 cd/m^2) defined the range as extending from 2.5 pps to 12.5 pps with a maximum disturbance at 6.5 pps, the source length-to-spacing (between centres) fraction having been kept constant at 0.25.

From these investigations it can be concluded that flicker is not disturbing under normal tunnel lighting conditions so long as the flicker frequency is less than 2.5 pps or greater than 15 pps. The relation between luminaire spacing and speed travelled needed to avoid this forbidden range of frequencies is given in figure 5.14.

With regard to the influence that the source-length/spacing fraction has on

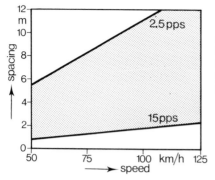

Figure 5.14 Forbidden luminaire spacing (shaded area) as a function of driver speed needed to avoid producing disturbing flicker in a tunnel (flicker frequency lower than 2.5 pps or greater than 15 pps). (CIE)

the severity of the flicker phenomenon at these frequencies, Schreuder (1964) also found clear evidence to suggest that the disturbance decreases when this fraction is made larger. Although no detailed quantitative data on this aspect of the problem were obtained, it does mean that in practice the danger of disturbance due to flicker can be reduced by making the bright area of the luminaire (or group of luminaires) in the lengthwise direction long compared with the dark area between them.

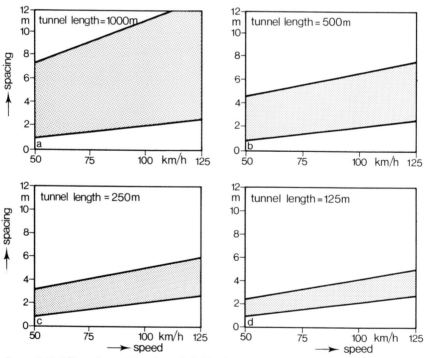

Figure 5.15 Effect of tunnel length on forbidden luminaire spacing with regard to disturbing flicker. (Walthert)

Finally, it has been found that the total exposure time to flicker has an influence on the range of frequencies over which the effect will prove troublesome: the shorter the exposure time (viz. the shorter the tunnel or the higher the driving speed), the smaller the range of disturbing frequencies becomes. Walthert (1977b) investigated the change in the lower limit of the forbidden frequency range with change in speed and tunnel length. The effect that this change has on the upper limit of the forbidden range of luminaire spacings, found by Walthert, is shown in figure 5.15, from which it can be seen that the critical frequency range decreases with decrease in tunnel length.

Part 2

EQUIPMENT

Chapter 6

Lamps

Electric lamps for normal lighting purposes may be divided into two main groups: incandescent and discharge. The light from an incandescent lamp is generated by bringing a filament to a high temperature by passing an electric current through it, whereas discharge lamps produce light by excitation of a gas or vapour contained between two electrodes. Discharge lamps may be further grouped according to whether the gas is contained at high or low pressure, with sub-divisions within each group according to the type of gas employed and the presence or absence of a fluorescent coating on the inner wall of the lamp envelope, it being the combination of these three factors that determines the type of light emitted.

The final lamp classification is shown in figure 6.1.

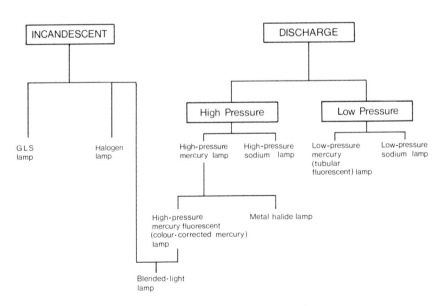

Figure 6.1 General classification of electric lamps.

6.1 Principal Characteristics

The principal characteristics of light sources are
(a) Luminous flux, expressed in lumens (lm)
(b) Luminous efficacy, expressed in lumens per watt (lm/W)
(c) Lamp mortality, usually expressed as the number of operating hours elapsed before a certain percentage of the lamps fail (See Chapter 11)
(d) Lumen depreciation, usually expressed using the lumen depreciation factor, this being the percentage of the original luminous flux after a certain number of operating hours (see Chapter 11)
(e) Luminance, expressed in candelas per square metre (cd/m^2) or per square centimetre (cd/cm^2)
(f) Colour appearance. The colour impression received when looking at the light source itself. The colour temperature (in Kelvin) can be used to indicate the colour appearance of a source. The lower the colour temperature, the more reddish the source; the higher the colour temperature, the more bluish the source.
(g) Colour rendering properties, for which there is the colour rendering index: R_a
(h) Physical size
(i) Permitted burning positions
(j) Warm-up time
(k) Re-ignition time
(l) Whether or not ballast and/or ignitor is needed.

A general survey of important characteristics of lamps that are representative of the groups shown in figure 6.1 is given in table 6.1.

In general, discharge lamps have a higher (sometimes much higher) efficacy than that of incandescent lamps and their life is considerably longer. Apart from certain exceptions, colour rendering is inferior to that given by incandescent lamps. Some form of external ballast is necessary, with the sole exception of the blended light lamp to which we shall refer later.

Unlike the incandescent lamp, most discharge lamps need a certain warm-up time before full light output is reached. Re-ignition immediately after switch-off is not instantaneous for all types. Finally, with some discharge lamps, the burning position of the lamp (viz. the orientation of the lamp axis) is important if proper performance is to be maintained.

6.2 Incandescent Lamps

Incandescent lamps form the oldest lamp family still in use and, basically because they are easy to work with and cheap, remain the mainstay of home

Table 6.1 Survey of characteristics of representative lamp types

Lamp type (I) clear bulb (II) bulb with diffusing layer (III) bulb with fluorescent layer		lamp wattage (W)	luminous efficacy (lm/W)		luminance (cd/cm²)	general lamp properties			
			lamp	lamp + ballast		colour rendering for road lighting application	colour rendering index	colour appearance	approximate luminous flux range (lm)
normal incandescent	(I)	100	14*	–	700	excellent	100	warm-white	100– 3 500
halogen incandescent	(I)	1000	22*	–	1000	excellent	100	warm-white	1000– 25000
tubular fluorescent 29	(III)	40	79	64	0.8	good	50	warm-white	100– 10000
tubular fluorescent 33	(III)	40	79	64	0.8	good	65	white	100– 10000
tubular fluorescent 34	(III)	40	51	41	0.5	excellent	85	white	200– 6000
tubular fluorescent 84	(III)	40	93	73	0.8	excellent	85	white	3000– 6000
high-pressure mercury	(I)	400	52	49	460	modest	20	blue-white	2000–125 000
high-pressure mercury	(III)	400	57	54	12	modest	40	white	2000–125 000
blended light	(III)	250	21	21**	5	good	60	warm-white	1000– 12000
metal halide	(I)	400	80	71	750	good	70	white	20000–200000
metal halide	(II)	400	75	67	15	good	70	white	20000–200000
high-pressure sodium	(I)	400	120	110	550	modest	25	yellow-white	3000–130000
high-pressure sodium	(II)	400	117	107	25	modest	25	yellow-white	3000–130000
low-pressure sodium	(I)	180	180	150	10	none	–	yellow	2000– 35000

* no ballast required
** built-in ballast

lighting, although their usefulness in many other fields has declined over the years with the development of the various types of discharge lamp. A relative newcomer to the family is the tungsten halogen incandescent lamp.

The incandescent lamp produces light with a continuous spectrum (see colour plate 1, facing page 96) by bringing a filament to incandescence. The limited life and efficacy of the normal incandescent lamp are the immediate outcome of evaporation of the tungsten filament caused by its high operating temperature, a process which, although slowed down by the presence of an inert gas, eventually leads to blackening of the bulb and finally to failure of the lamp.

Tungsten halogen incandescent lamps were introduced in about 1960. In these lamps the blackening of the bulb by evaporation of the filament is prevented by adding a halogen to the fill gas. Use of this principle makes it necessary to adopt a smaller bulb and to increase the pressure of the inert gas, but the temperature of the filament can then be higher than with normal incandescent lamps resulting in a considerably higher luminous efficacy at the same or longer lamp life.

These features of the tungsten halogen lamp, together with its higher luminance, are well illustrated by the extensive use made of these lamps in floodlighting, motor car lights, traffic signs, projectors and spotlights. Nevertheless, neither these nor normal incandescent lamps are to be recommended for road lighting purposes, chiefly on account of their low efficacy and rather limited life compared with other lamps available.

6.3 Tubular Fluorescent Lamps

The tubular fluorescent lamp (figure 6.2) consists of a tubular bulb having an electrode sealed into each end and containing mercury vapour at low pressure with a small amount of inert gas to aid starting. The inner surface of the tube is coated with fluorescent powders. When a current is passed through this gas mixture, predominantly ultraviolet radiation is produced (figure 6.3). The fluorescent powders transform this radiation into visible light.

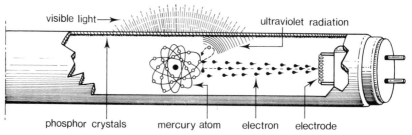

Figure 6.2 Part-section of a tubular fluorescent lamp, illustrating the mechanism of discharge.

By varying the composition of the fluorescent powders, lamp types with different colour appearance and colour rendering properties can be made. The different types can be denoted by the use of a numerical colour designation suffix. Where fluorescent lamps are used in road lighting, an area where colour rendering is not as important as high efficacy, the best choice would usually be the colour 29 lamp (warm-white) or the colour 33 (white) which have colour rendering indices (or R_a values) of 50 and 65 respectively and an efficacy of about 80 lm/W (colour plate 1, facing page 96). Fluorescent lamps with higher colour rendering indices but lower efficacies or those with higher colour rendering indices but costing more (such as the new-generation lamps: 83, 84 and 86) are not recommended for use in road lighting.

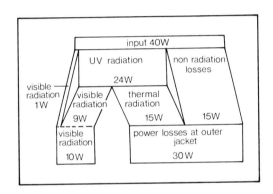

Figure 6.3 Diagram representing schematically the conversion of power in a tubular fluorescent lamp.

Fluorescent lamps are commonly available in the range of about 100 lm to 10 000 lm (corresponding to 4 W to 110 W).

The length of a given lamp is in general related to its wattage, the higher the wattage, the longer the tube. This, and their rather low luminous flux compared with other discharge lamps are decided disadvantages in so far as the general use of these lamps in road lighting is concerned. Their length makes good optical control of light output in any plane through the lamp axis impossible to achieve. Furthermore, where moderate to high lighting levels are required, the luminaire would need to house several lamps in order to produce the luminous flux needed, and this would result in the luminaire having a low light output ratio.

However, where the application calls for optical control in the plane perpendicular to the lamp axis rather than parallel to it – as in tunnel lighting – the fluorescent lamp can become a practical proposition. Also, applications where little optical control is called for and low lighting levels suffice, as for example in the lighting of certain residential and pedestrian areas, the fluorescent lamp can be used to advantage; although here the lamp used is generally circular or U-shaped, shapes specially suited for use in the post-top luminaires designed for use in such areas.

Fluorescent lamps should be used with caution in those areas where the air temperature is subject to large drops, since the luminous flux of these lamps decreases from its maximum value with fall in ambient temperature. (This problem can usually be avoided by operating the lamps in luminaires of the closed variety.)

In short, road lighting installations employing tubular fluorescent lamps usually do not offer an economical solution compared with installations using other discharge lamps.

6.4 High Pressure Mercury Vapour Lamps

As with all discharge lamps, high-pressure mercury lamps need a ballast in order to operate (See Sec. 6.8). Unlike other discharge lamps, however, the high-pressure mercury lamp is available in a special version in which the ballast is built into the lamp itself. This so-called blended-light lamp, although less efficient in operation than the normal high-pressure mercury lamp, can thus be connected direct to the mains just like an incandescent lamp.

6.4.1 Normal High Pressure Mercury Vapour Lamps (HP, HPL-N)

Under normal conditions of operation the gas atmosphere in the discharge tube of a high-pressure mercury vapour lamp (figure 6.4) consists of vaporised mercury at a pressure of between 0.2 MPa and 1 MPa, with a small quantity of inert gas to facilitate ignition.

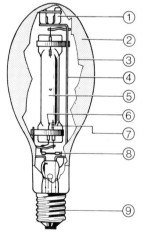

Figure 6.4 High-pressure mercury vapour lamp. 1. Support spring 2. Ovoid hard-glass outer envelope 3. Inner phosphor coating 4. Lead-in wire/support 5. Quartz discharge tube 6. Auxiliary electrode 7. Main electrode 8. Starting resistor 9. Screw base.

The discharge tube, which is of quartz to withstand the high operating temperature, has a main electrode sealed into each end. Adjacent to one or both of these electrodes is an auxiliary starting electrode. The glass outer bulb normally contains an inert gas (at atmospheric pressure when the lamp is operating) which, in addition to protecting the discharge tube and its lead-in wires from atmospheric influences at the high operating temperatures prevailing, stabilises the arc discharge by maintaining a near constant temperature over the normal range of ambient conditions.

High-pressure mercury vapour lamps are of basically two types: clear glass (HP) and phosphor coated (HPL-N).

Clear-glass lamps. These are characterised, when burning, by their bluish-white colour appearance. The arc in fact produces a line spectrum (see colour plate 1, facing page 96) with emission within the visible region at the blue, green and yellow wavelengths, there being an absence of red radiation. Their luminous efficacy ranges from 40 lm/W for an 80 watt lamp to more than 50 lm/W for one of 400 watts.

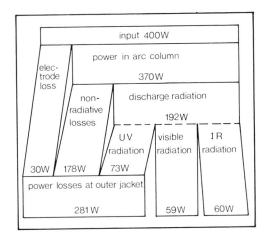

Figure 6.5 Power conversion in a high-pressure mercury vapour lamp.

Phosphor-coated lamps. The pure mercury arc has both poor colour appearance and colour rendering but emits a significant portion of its energy in the ultraviolet region of the spectrum (see figure 6.5), viz. between 100 nm and 400 nm. By the use of a phosphor coating on the inside of the outer envelope, as is done in the fluorescent versions of this lamp, this ultraviolet energy can be made to introduce a red component (see again colour plate 1) which greatly improves the lamp's colour appearance and colour rendering, the index for the latter rising from approximately 20 to 40. The improvement in efficacy is only slight (about 10 per cent) as the eye's sensitivity to red is low and, in addition, some of the visible radiation generated by the discharge is absorbed by the phosphor coating.

The main advantages of the mercury vapour discharge lamp are its high reliability and high efficacy compared with incandescent lamps and its moderate price in comparison with other high-pressure discharge lamps. It is the traditional lamp for road lighting, although with rising energy prices the higher efficacy of the sodium discharge lamps (see Sec. 6.6 and 6.7) serves to tip the balance in their favour when white light, as given by the high-pressure mercury vapour lamp, is not needed.

High-pressure mercury vapour lamps are available in the approximate range 2000 lm to 125000 lm (corresponding to 50 W to 2000 W).

6.4.2 Blended Light Lamps (MLL)

The blended-light lamp (figure 6.6) consists of a gas-filled bulb coated on its inside with a phosphor and containing a normal mercury discharge tube which is connected in series with a tungsten filament. The filament acts as a ballast for the discharge, so stabilising the lamp current. No external ballast is needed.

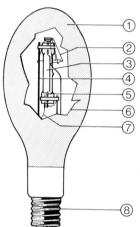

Figure 6.6 Blended-light lamp. 1. Ovoid hard-glass outer envelope 2. Coiled filament 3. Quartz discharge tube 4. Support 5. Main electrode 6. Internal phosphor coating 7. Lead-in wire 8. Screw base.

The blended-light lamp, like the HPL-N phosphor-coated lamp from which it is derived, emits visible radiation from the mercury discharge tube plus that obtained by conversion of the ultraviolet radiation at the phosphor coating. The incandescent filament of the blended-light lamp adds its own warm-coloured light to this visible radiation (colour plate 1) to give the lamp a more pleasant colour appearance.

Blended-light lamps can, by virtue of the built-in ballast, be connected direct to the mains. This means that any existing installation employing incandescent lamps can be easily modernised using blended-light lamps, which have almost twice the efficacy and five times the operating life, at no extra cost in terms of special gear, wiring or luminaires. However, compared with HPL lamps they have a lower efficacy and a shorter life.

6.5 Metal Halide Lamps (HPI, HPI/T)

Metal halide lamps (figure 6.7) are similar in construction to the high-pressure mercury vapour lamp, the major difference being that the discharge tube of the former contains one or more metal halides in addition to mercury. These halides are partly vaporised when the lamp reaches its normal operating temperature. The halide vapour is then dissociated in the hot central region of the arc into the halogen and the metal, with the vaporised metal radiating its appropriate spectrum.

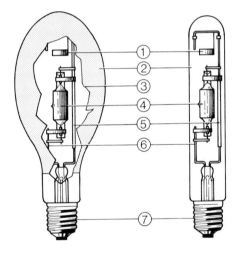

Figure 6.7 Metal halide lamps are available in two versions: diffuse ovoid and clear tubular. 1. Getter ring for maintaining high vacuum 2. Hard-glass outer bulb 3. Internal phosphor coating 4. Quartz discharge tube 5. Sleeve protecting the support 6. Lead-in wire/support 7. Screw base.

A number of different metal halide lamps have been developed. A typical combination of halides used is that comprising the iodides of sodium, indium and thallium. These halides give an increase in intensity in three spectral bands: blue, green and yellow-red (colour plate 1, facing page 96). Colour rendering is improved in comparison with that of the high-pressure mercury lamp. Efficacy is also increased considerably; this is because the radiation emitted lies in the region of the spectrum to which the eyes are highly sensitive. The result is a lamp with reasonable colour rendering (an R_a of 70) and an efficacy of over 80 lm/W for a 400 watt lamp.

Its application is floodlighting and sports-field lighting, there being an essential need here for a compact (thus easily focussed), efficient source of white light. It is also being used more and more in industrial lighting. The life of the metal halide lamps is for the time being at least, shorter than that of other discharge lamps, so that this lamp does not as yet often offer an economical solution for road lighting.

Most metal-halide lamps are commonly available in the approximate range 20000 lm to 200000 lm (corresponding to 250 W to 2000 W).

6.6 High Pressure Sodium Lamps (SON, SON/T, SON/H)

A more recently developed discharge lamp having about twice the efficacy of the high-pressure mercury vapour lamp is the high-pressure sodium lamp (figure 6.8).

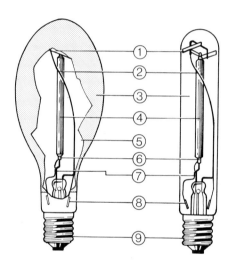

Figure 6.8 High-pressure sodium lamps are available in two versions: diffuse ovoid and clear tubular. 1. Support springs to maintain discharge tube alignment 2. Lead-in wire 3. Hard-glass outer bulb 4. Translucent aluminium oxide discharge tube 5. Inner diffusing coating 6. End cap of discharge tube 7. Lead-in wire/support 8. Getter rings for maintaining high vacuum 9. Screw base.

Under normal conditions of operation the gas atmosphere of the discharge tube consists of a mixture of vaporised sodium and mercury at a pressure of between 13 kPa and 26 kPa, with a small quantity of inert gas to facilitate ignition.

It has been known for a long time that sodium gives a higher proportion of radiation at wavelengths in the visible range for which the eye has its maximum sensitivity than does mercury, hence the higher efficacy of the sodium lamp (figure 6.9). However, the presence of sodium in the discharge tube at the relatively high temperatures needed in a high-pressure discharge necessitates the use of a light-transmitting wall material resistant to sodium. Quartz, which is employed as a wall material for the discharge tube of the

90

high-pressure mercury vapour lamp, is affected by sodium at these high temperatures and so cannot be used. The hunt for a suitable material ended with the discovery of sintered alumina, a translucent material resistant to attack by sodium and one able to transmit visible radiation very well – it has a transmission of more than 90 per cent.

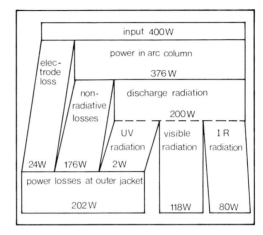

Figure 6.9 Power conversion in a high-pressure sodium lamp.

The colour rendering and colour appearance of high-pressure sodium lamps are more than adequate for the majority of road lighting applications; indeed, lighting using these lamps is often considered more pleasant than mercury lighting. This, together with the higher efficacy of high-pressure sodium lighting may be expected to weigh heavily in its favour for many future outdoor lighting applications in general.

High-pressure sodium lamps are available with two types of outer bulb: clear-glass tubular (SON/T) and ovoid (SON). The ovoid version has the same shape as the HPL-N (high-pressure mercury) lamp. The bulb is coated on the inside with a diffusing powder which serves to enlarge the light-emitting area, so decreasing the lamp's luminance.

High-pressure sodium lamps are available in the approximate range 3000 lm to 130000 lm (50 W to 1000 W).

A special version of the high-pressure sodium lamp, the SON/H lamp, has been developed that can be used to replace the high-pressure mercury lamp in existing installations without the necessity of changing the ballast or adding a starter system (see Sec. 6.8.2 Starters and Ignitors). The SON/H lamp consumes less energy than the mercury lamp it replaces (up to 15 per cent less) for a higher luminous flux (up to 40 per cent higher). Its efficacy is, however, not so high as that of the standard high-pressure sodium lamp, and for this reason it is not recommended for use in new installations.

6.7 Low Pressure Sodium Lamps (SOX)

The U-shaped soda-lime glass discharge tube of the low-pressure sodium lamp (figure 6.10) has a layer of sodium-resistant borate glass applied to its inside wall. Under normal conditions of operation the gas atmosphere of the discharge tube consists of vaporised sodium, at a pressure of about 300 mPa, and a mixture of inert gases to obtain a low ignition voltage.

Prior to ignition, the sodium is dispersed in dimples formed in the wall of the discharge tube. After ignition, the discharge first takes place through the inert gases. As the temperature in the tube gradually increases, some of the sodium vaporises and takes over the discharge. At switch-off, the sodium condenses and again collects at the dimples, these being the coolest spots in the tube.

The discharge tube is contained in an evacuated tubular glass envelope coated on its inner surface with indium oxide. This coating acts as an infrared

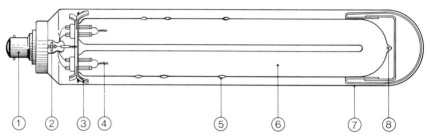

Figure 6.10 Low-pressure sodium lamp. 1. Bayonet cap (for accurate positioning) 2. Getter for maintaining high vacuum 3. Support springs 4. Electrodes 5. Sodium retaining dimples 6. Glass discharge tube 7. Outer envelope coated with infrared reflector 8. Exhaust pip.

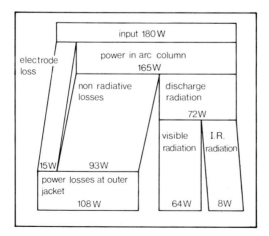

Figure 6.11 Power conversion in a low-pressure sodium lamp.

92

reflector and so helps to maintain the wall of the discharge tube at the operating temperature of 270 °C, the temperature at which the lamp's optimum vapour pressure is reached.

The low-pressure sodium lamp emits monochromatic radiation (see colour plate 1, facing page 96). Colour rendering is consequently non-existent, but the wavelength of this radiation is close to that for which the eye has its maximum sensitivity and it is this that gives the lamp its extremely high luminous efficacy (figure 6.11) of up to about 200 lm/W. The low-pressure sodium lamp has, in fact, the highest efficacy of all lamp types. It therefore finds application where efficacy is all important and where good contrast recognition and visual acuity count for more than good colour rendering, as is the case with many road lighting applications.

Low-pressure sodium lamps are available in the approximate range 2000 lm to 35000 lm (18 W to 180 W).

6.8 Control Gear

6.8.1 Ballasts

All discharge lamps need a series impedance to limit the lamp current. Were such a device not used, there would be nothing to prevent this current from increasing to the point where lamp destruction takes place. Such an impedance, or ballast as it is called, forms part of the control gear necessary for the operation of these lamps.

Apart from providing good stabilisation of the lamp current, the ballast must:

(a) Have a high power factor so as to ensure economic use of the supply system
(b) Limit the generation of harmonics
(c) Present a high impedance to frequencies used for switching purposes
(d) Offer adequate suppression of any radio interference that might be produced by the lamp
(e) In many cases, furnish the correct starting conditions for the lamp concerned.

Another group of requirements is constituted by the wishes of both luminaire manufacturer and user to have available ballasts of small dimensions, low losses, long life and low hum level.

An effective form of ballast is the inductance, or choke, placed in series with the lamp (figure 6.12a). Inherently, the power factor of this circuit is low, viz. about 0.5 lagging, but this can be increased to 0.85 or greater by simply connecting a capacitor in shunt across the a.c. supply (figure 6.12b). To correct the power factor to unity by this method is not possible, because of the waveform distortion produced by discharge lamps.

A draw-back of the shunt-connected capacitor is that it presents a low impedance to the high-frequency signals sometimes used for the remote switching of road lighting installations: the capacitors in the installation would therefore short-circuit the distribution network for these signals. This problem can be simply solved, however, by connecting a small inductance, or filter coil, in series with the shunt capacitor to give the necessary value of circuit impedance.

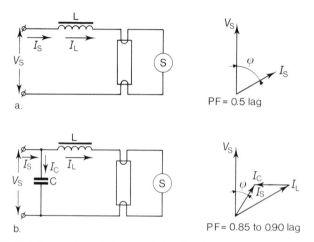

Figure 6.12 Switch-start lamp ballasts (a) Simple choke ballast (b) Choke ballast with parallel connected compensating capacitor.

Power factor correction obtained by connecting a capacitor in series with the inductive ballast rather than across the a.c. supply is not often used. Although amenable to high-frequency remote control switching, it approximately doubles the warm-up time of the lamp and calls for the use of high-value, and thus physically large, capacitors.

6.8.2 Starters and Ignitors

All discharge lamps used in road lighting, with the exception of the high-pressure mercury lamp, need a voltage higher than that of the mains supply to initiate the arc discharge. Such lamps must therefore be operated in conjunction with some form of starting device. This device may constitute a separate item of control gear, it may form an integral part of the ballast, or it may be built into the lamp itself, depending on the lamp type concerned.

Fluorescent lamps. Tubular fluorescent lamps are of two types, switch-start and starterless. Switch-start lamps are started by heating the lamp electrodes before application of the high starting voltage. This preheating, which may take a few seconds, is initiated by a separate bimetal starter switch.

Starterless fluorescent lamps are further sub-divided into rapid start and instant start types. The electrodes of a rapid-start lamp are heated continuously, from the moment of switch-on, from low-voltage windings built into the ballast. Since the ignition 'kick' supplied by the action of the starter is not available, these lamps are provided with an external ignition strip to aid starting. In an instant-start fluorescent lamp, ignition is dependent solely on the application of a high voltage across the lamp, this voltage being supplied by the ballast.

All starterless lamps are given a transparent, water-repellent coating on the outside of the tube for reliable starting under humid conditions. The efficacy of these lamps is in general lower than that of switch-start lamps of the same power rating. The feature they offer in connection with outdoor applications such as road lighting, however, is that they give reliable starting at very low temperatures, down to $-15°C$ compared to the $+5°C$ minimum of the switch-start lamps.

Metal halide and low-pressure sodium lamps. The peak voltages needed for the ignition of these lamps are well above the voltage of the normal mains supply, viz. 600–700 volts for the metal halide lamp and 400–600 volts for the low-pressure sodium lamp. Both these lamps can be started using an electronic thyristor ignitor connected across the lamp electrodes. The ignitor generates a series of high-voltage pulses of the required magnitude, and an electronic circuit takes care that these pulses cease after ignition has taken place.

Since the ignition voltage of the low-pressure sodium lamp is relatively low, ignition can also be achieved with the aid of an auto-leak transformer, which at the same time functions as an inductive ballast. Until recently, in fact, this method of ignition was obligatory, since with all low-pressure sodium lamps the operating voltage can, during warm-up, momentarily reach high values – the high-wattage lamps even have operating voltages permanently higher than that of the mains supply. However, by transforming the normally sinusoidal lamp current into a waveform resembling a square wave, as is currently the practice, the operating voltage of the lamp can be decreased to below mains voltage. This permits the use of an electronic ignitor with separate ballast, the electronic squaring circuit normally being an integral part of the former. The advantage of this ignitor/ballast combination, sometimes referred to as a hybrid ballast, compared with the auto-leak transformer is that it considerably decreases the wattage loss, weight, and size of the control gear.

High-pressure sodium lamps. These lamps need a peak voltage of some 3000 volts for ignition. Because of this, the ignitor used is not connected across the lamp direct, but via a tapping on the choke winding, which acts as a step-up transformer for the starting pulses from the ignitor.

Some high-pressure sodium lamps have the ignition system built into the lamp itself. Each lamp is then equipped with a bimetal switch, the precise function of which differs from manufacturer to manufacturer.

Depending on type, some of these high-pressure sodium lamps with built-in ignitor can be used in conjunction with normal high-pressure mercury lamp ballasts, thereby allowing older high-pressure mercury installations to be updated with no expenditure on control gear (Sec. 6.6).

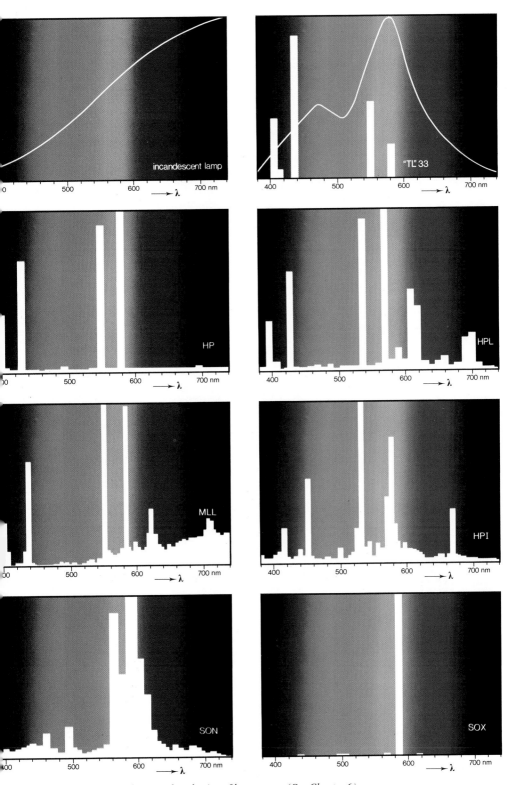

Relative spectral energy distributions of a selection of lamp types. (See Chapter 6)

2 *Open-air road lighting laboratory near Eindhoven, the Netherlands. (See Sec. 14.2.3)*

Chapter 7

Luminaires

A luminaire is a device that controls the distribution of the light given by a lamp or lamps and which includes all the items necessary for fixing and protecting those lamps and for connecting them to the supply circuit.

The luminaires employed in road lighting are of three basic types: conventional luminaires, catenary luminaires and floodlights. Conventional luminaires are all those designed for column, wall, or spanwire mounting such that the main vertical plane of symmetry lies at right angles to the axis of the road thus throwing the main part of their light along the road. This is in contrast to catenary luminaires, which are designed to be suspended from a cable (catenary) with the main plane of symmetry parallel to the axis of the road thus throwing the main part of their light across the road. Finally, there is the floodlight. Unlike conventional and catenary luminaires, the floodlight is completely free as far as any fixed orientation with respect to the road is concerned.

7.1 Principal Characteristics

The principal characteristics of the various luminaires employed in road lighting can be listed under the following headings

(a) Photometric: luminous intensity distribution and light output ratio; luminaire luminance
(b) Thermal: heat resistance; operating temperature within the luminaire
(c) Mechanical and aerodynamic: sturdiness; toughness (degree of vandal proofing); weight, size and shape
(d) Electrical: nature of control-gear housing; electrical connections; safety
(e) Installation and maintenance: ergonomic construction; accessibility of lamps and control gear; cleanability
(f) Resistance to atmospheric attack and pollution: corrosion resistance; resistance to ultraviolet and visible radiations; resistance to ingress of dust and dirt
(g) Type of mounting: post top, bracket, or cable suspended
(h) Aesthetics: general appearance and styling.

7.1.1 Photometric Characteristics

Light distribution and light output ratio

The desired luminous intensity or light distribution of a luminaire is achieved through the application of one or more of the physical phenomena: reflection, refraction, and diffuse transmission, although the last mentioned is normally only employed where optical control is not critical, as in many of the more decorative luminaires designed for use in pedestrian areas. Most road lighting luminaires also make use of shielding in one form or another, principally to obtain the required degree of glare control. The shielding function may be performed by mirror reflectors, by white-painted surfaces or, where very stringent glare control is required, by mat-black surfaces. Typical techniques employed to control light distribution are illustrated in figure 7.1.

The way in which these various control techniques are employed in a given luminaire, and the optical properties of the materials used will combine to determine the light output ratio of the luminaire. This is defined as the ratio of the light output of the luminaire to the sum of the light outputs of the lamps it contains. Of particular concern in road lighting is a luminaire's downward light output ratio (d.l.o.r.), viz. the flux emitted below the horizontal (and thus in the general direction of the road) as a fraction of total lamp flux. The d.l.o.r. is thus a measure of luminaire efficiency: the higher the ratio, the higher the efficiency.

a. reflection

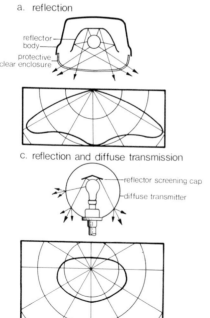

b. mostly refraction

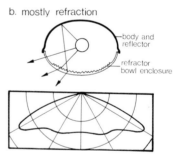

c. reflection and diffuse transmission

Figure 7.1 Control of luminaire light output by means of (a) reflection (b) refraction, and (c) reflection and diffuse transmission, with resulting luminous intensity diagrams.

Luminaire luminance

The size and luminance of the lamp have an important bearing on the composition, dimensions and shape required for the optical system. For example, the mirror reflector system employed with a lamp having a small light-emitting area (such as the SON/T and HPI/T lamps) can be kept relatively small whilst still producing a well-controlled light distribution. Of course, such a small luminaire will have a relatively high luminance, so that, in order to prevent glare, special attention has to be given to the light distribution. With small lamps, any irregularity in the mirror will show up as a discontinuity in the luminous intensity distribution and consequently as a patch of different brightness on the surface of the road. The mirror is therefore sometimes given a faceted or a hammered-finish, or is used in combination with a diffusing panel placed near the lamp to overcome this difficulty.

Lamps with larger light-emitting areas call for correspondingly larger optical systems. However, the lower luminaire luminances associated with these systems means that glare can more easily be restricted and small irregularities in the optical system will not cause disturbing irregularities in the luminance pattern on the road surface. Tolerances in lamp dimensions and in lamp positioning can also be greater without severely disturbing the light distribution.

7.1.2 Thermal Characteristics

Heat resistance

The materials used within the body of the luminaire and the body materials themselves should be capable of withstanding the heat produced by the lamp used in it.

Operating temperature

The temperature should not rise so high during operation as to upset the correct functioning of the lamp or lamps. The volume of a luminaire, especially if it is of the totally enclosed type, is important in this respect, and in some small luminaires fins on the housing are needed to promote cooling, especially where high-wattage sources are employed.

7.1.3 Mechanical and Aerodynamic Characteristics

Sturdiness

The luminaire and its mounting attachments should be of sturdy construction to ensure a good, steady positioning of the luminaire and its contents. Any weakness here could lead to changes in the luminaire's light distribution and consequent changes in the planned lighting quality of the installation as a whole.

Toughness
Luminaires mounted low down in residential areas are especially vulnerable to attack by vandals. They should therefore be constructed of tough materials as far as possible.

Weight, size, and shape
The weight, size, and shape of a luminaire determine the strength of the mast or cable needed to support it and withstand windforces and vibrations caused by air currents and passing road traffic. The demands placed on mast strength will be lowest for a light luminaire that presents a small, streamlined area to the wind, the streamlining of a luminaire being denoted by its 'shape factor'.

7.1.4 Electrical Characteristics

The construction of a luminaire should be such as to render it safe electrically to all those involved in its handling. The International Electrotechnical Commission (IEC) and the International Commission on Rules for the Approval of Electrical Equipment (CEE) classify luminaires according to the degree of protection afforded against electrical shock (table 7.1); those described as class 0 (no provision for earthing) are not suitable for road-lighting purposes and should therefore not be used.

Table 7.1 IEC/CEE classification of luminaires according to type of electrical protection

Luminaire class	Electrical protection
0	A luminaire having functional insulation, but not double insulation or reinforced insulation throughout, and without provision for earthing.
I	A luminaire having at least functional insulation throughout and provided with an earthing terminal or earthing contact, and, for luminaires designed for connection by means of a flexible cable or cord, provided with either an appliance inlet with earthing contact, or a non-detachable flexible cable or cord with earthing conductor and a plug with earthing contact.
II	A luminaire with double insulation and/or reinforced insulation throughout and without provision for earthing.
III	A luminaire designed for connection to extra-low-voltage circuits, and which has no circuits, either internal or external, which operate at a voltage other than extra-low safety voltage.

Minimum sizes of the electrical conductors (wires and connecting blocks) used in the luminaire should be suitable for the actual electrical load. The cable insulation must be sufficient for the high ignition voltages (which may be far higher than the operating voltage, Chapter 6) and must withstand the often high temperature in the luminaire during operation. To relieve possible strain on the cables and their connections a means of clamping the cables is often required.

7.1.5 Installation and Maintenance Characteristics

Ergonomic design
Mounting, relamping, and cleaning must usually be carried out high above ground level, so the ergonomic design of the luminaire should be such as to make these operations as easy as possible to perform; covers, for example, should be hinged so that the electrician has his hands free to work on the lamp and ballast and to make or check the various electrical connections.

Accessibility and cleanability
A luminaire of simple construction with clean lines will, by improving accessibility and facilitating cleaning, not only help to shorten relamping and cleaning times (thereby reducing maintenance costs), it will also reduce the time a road or part of a road is out of commission whilst maintenance of the lighting installation is in progress.

7.1.6 Resistance to Atmospheric Attack and Pollution

Corrosion, dirt, and moisture-resistance
The atmosphere can contain many potentially corrosive gases which, in the presence of moisture vapour, will form highly corrosive compounds. Luminaires made from corrosion-resistant materials or having protective finishes should therefore be chosen for use in all areas where this danger is known to exist.
Closed luminaires afford much better protection against corrosion and dirt accumulation than do open luminaires. However, due to the variation in temperature between the air inside and that outside a luminaire after switching on or off, pressure differences across the luminaire's cover-seal are bound to occur. The seal should therefore be able to 'breathe', whilst retaining its dust and moisture proof properties, otherwise the corrosive gases, moisture, and dirt will be sucked into the luminaire during the cooling off period thereby giving rise to corrosion and fouling of the possibly untreated, or unprotected, inner surfaces.

Radiation resistance

Any plastics materials used in the fabrication of a luminaire should be impervious to ultraviolet and visible radiations. It is especially important that no discolouration of the outer body of the luminaire should occur.

7.1.7 Aesthetics

No less important than the functional characteristics of a luminaire is what may be termed its aesthetic appeal, that is to say its general appearance and styling. It must be remembered that during the hours of daylight, while the lighting installation is not in use, it will be clearly visible to all. In pedestrian areas especially, even a dormant installation can make a positive contribution to the attractiveness of the locality. But no matter what the type of installation, it must at all times harmonise as far as possible with its surroundings.

7.2 Conventional and Catenary Luminaires

7.2.1 Photometric Classification

In order to be able to indicate the suitability or otherwise of a luminaire for a given application, some form of system that classifies luminaires according to their photometric characteristics is needed. Several such classification systems have been formulated, and these are outlined in this section.

Classification system of the International Commission on Illumination (CIE)
The CIE in its Publication No. 34 (1976a) entitled 'Road Lighting Lantern and Installation Data – Photometrics, Classification and Performance', proposes a luminaire classification based upon three features of photometric performance:

(a) The extent to which the light from the luminaire is 'thrown' in the lengthwise direction of the road, called the 'throw' of the luminaire
(b) The extent to which the light is 'spread out' across the road, called the 'spread' of the luminaire
(c) The extent of the facility for controlling glare, called the 'control' of the luminaire.

The *throw* is defined by the angle (γ_{max}) that the beam axis makes with the downward vertical. The beam axis is defined by the direction midway between the two directions of 90 % I_{max} in the vertical plane of maximum intensity (see figure 7.2), Throw is illustrated in figure 7.3.

The *spread* is determined by the position of the line running parallel to the road axis that just touches the far side of the 90 per cent I_{max} contour on the

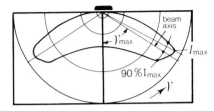

Figure 7.2 Luminous intensity distribution of a road lighting luminaire in the plane of maximum intensity (I_{max}) showing the 'beam axis', making an angle γ_{max} with the downward vertical, located midway between the two directions of $90\%\ I_{max}$. The angle γ_{max} defines the 'throw' of the luminaire (see also figure 7.3).

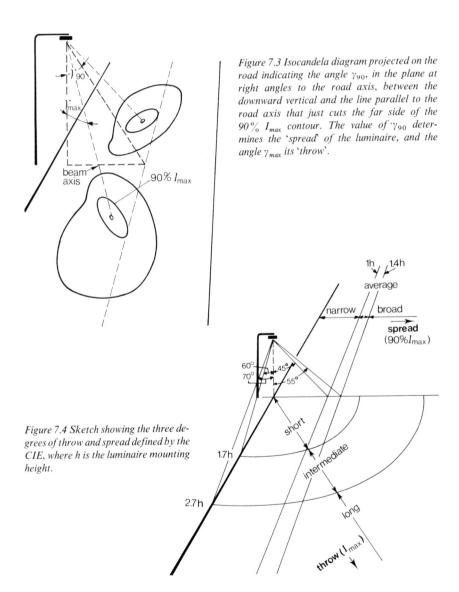

Figure 7.3 Isocandela diagram projected on the road indicating the angle γ_{90}, in the plane at right angles to the road axis, between the downward vertical and the line parallel to the road axis that just cuts the far side of the $90\%\ I_{max}$ contour. The value of γ_{90} determines the 'spread' of the luminaire, and the angle γ_{max} its 'throw'.

Figure 7.4 Sketch showing the three degrees of throw and spread defined by the CIE, where h is the luminaire mounting height.

103

road surface, figure 7.3. The position of this line is defined by the angle γ_{90}. The *control* is determined by those luminaire characteristics that also determine its glare control mark, G. Control is therefore defined by the specific luminaire index, *SLI* (Sec. 1.3.3).

The CIE recognises three degrees of throw, spread, and control. These are defined in table 7.2 together with the corresponding names. Figure 7.4 indicates, on a plan of the road, the coverage given by the three degrees of throw and spread in terms of luminaire mounting height h.

Table 7.2 Definition of the CIE classification system for the photometric properties of luminaires

Throw		Spread		Control	
Short	$\gamma_{max} < 60°$	narrow	$\gamma_{90} < 45°$	limited	SLI < 2
Inter-mediate	$60° \leqslant \gamma_{max} \leqslant 70°$	average	$45° \leqslant \gamma_{90} \leqslant 55°$	moderate	$2 \leqslant$ SLI $\leqslant 4$
Long	$\gamma_{max} > 70°$	broad	$\gamma_{90} > 55°$	tight	SLI > 4

The throw and spread of a luminaire are of course most easily determined from an isocandela diagram in which the isocandela contours are projected on the plane illuminated by the luminaire. If such a diagram is not available, they can also be determined from the isocandela diagram in zenithal projection (see Sec. 7.2.2).

It should be emphasised on passing that in the CIE definition of throw the angle determining the degree of throw lies in the plane of maximum intensity, that is to say in the vertical plane containing the beam axis of the luminaire. The definition says nothing about the angle made between this plane and the road axis, and this means that luminaires for which the plane of maximum intensity is more across the road than along it may, in practice, need to be spaced closer together, for the same angle of throw, than luminaires for which the angle between this plane and the road axis is more acute. For this reason, it would perhaps have been better had throw been defined not by the angle that the beam axis makes with the downward vertical, but by the angle that the projection of this axis on the vertical plane parallel to the road axis makes with the downward vertical.

In the first edition of 'International Recommendations for the Lighting of Public Thoroughfares' (CIE, 1965) the terms 'cut-off', 'semi-cut-off', and 'non-cut-off' were used to describe three types of luminaire luminous intensity distribution. This system of luminaire classification was used as a means of describing indirectly the degree of glare control possessed by an installation as a whole. In the updated CIE road-lighting recommendations, however, glare control is more properly given in terms of those parameters that directly determine the degree of glare likely to be experienced. The terms cut-off, semi-cut-off, and non-cut-off are thus no longer needed, but since they are nevertheless still used in various national road-lighting recommendations, they have been defined in table 7.3.

Table 7.3 Definition of the types of luminous intensity distributions according to the 1965 CIE Recommendations for the Lighting of Public Thoroughfares

	Maximum permissible value of intensity emitted at an elevation angle of		Direction of maximum intensity smaller than
	80°	90°	
Cut-off	30 cd/1000 lm	10 cd/1000 lm*	65°
Semi-cut-off	100 cd/1000 lm	50 cd/1000 lm*	75°
Non-cut-off	any	*	–

* Up to an absolute maximum value of 1000 cd.

Classification system of the Illuminating Engineering Society (IES) of North America (AINSI/IES, 1977)
The IES classification system is based upon features of photometric performance similar to those used by the CIE, namely the luminaire's longitudinal and lateral light distributions and the degree of glare control present. Again, like the CIE, the IES recognises three types of longitudinal distribution: short, medium, and long, but the IES defines the longitudinal distribution according to the highest intensity in the plane parallel to the road axis. Four lateral light distributions for conventional luminaires are defined by the IES on the basis of the position of the 50 % I_{max} contour instead of the 90 % I_{max} contour used by CIE, and these are numbered I to IV, figure 7.5. The extent of the glare-control facility is defined in terms of the luminous intensities at 80° and 90° above nadir, three degrees of control being recognised, as described by the terms cut-off, semi-cut-off, and non-cut-off, table 7.4.

Table 7.4 Definition of the degrees of glare control of luminaire light distribution – according to the IES of North America

	Maximum permissible value of intensity emitted at an elevation angle of	
	80°	90°
Cut-off	100 cd/1000 lm	25 cd/1000 lm
Semi-cut-off	200 cd/1000 lm	50 cd/1000 lm
Non-cut-off	any	any

It will be noticed, comparing tables 7.3 and 7.4, that there is little correspondence between this glare-control classification and that put forward by the CIE in 1965, although the same descriptive terms are used. Each has been evolved to fit in with the lighting recommendations promulgated by the body concerned, and special care must therefore be taken to ensure that the appropriate definition is used.

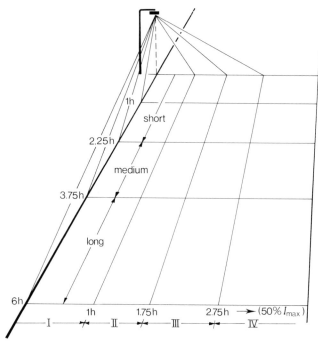

Figure 7.5 Sketch showing the three degrees of longitudinal and four degrees of lateral light distribution defined by the IES of North America, where h is the luminaire mounting height. The former are defined by the position of the I_{max} vector, and the latter by the position of the 50% I_{max} contour.

Classification system of the British Standards Institution (BSI, 1976)
The BSI classification system (table 7.5) is intended to permit the use of a general 'recepy' method for road-lighting design, the idea being that all the necessary information on photometric performance is contained in the classification system itself; so that for design purposes, not the individual luminous intensity distribution is required but only its classification (see Sec. 9.1.2).
It is perhaps unfortunate that the terms cut-off and semi-cut-off used by the BSI for the two classes of luminaires are the very same as those used by CIE (1965) and IES to describe the degree of glare control possessed by a luminaire, because in the BSI classification system the terms describe more than just glare control.

Table 7.5 Classification system for luminaires of the British Standards Institution

	In plane of principal vertical polar curve[1]		Limits of intensity within the cone from downward vertical to 30° therefrom cd/1000 lm		In vertical plane parallel to road axis		Intensity at 90° elevation cd/1000 lm
	Angle of elevation contained within the beam	Limits of peak intensity[2] cd/1000 lm			Elevation at which an intensity of 130 cd/1000 lm occurs		
		min. max.	min.	max.	min. max.		max.
Cut-off	65°	200 500	30	250[3]	72° 78°		15
Semi-cut-off	75°	180 500	30	250[3]	78° 84°		75

[1] The polar curve in the vertical plane taken halfway between the extremes in azimuth of the solid angle which is bounded by directions having intensities equal to 90% of the maximum intensity.
[2] The maximum intensity in the plane of the principal vertical polar curve as defined in footnote 1.
[3] Maximum not to exceed 80% of peak intensity.

7.2.2 Photometric Documentation

Before a lighting engineer can decide what luminaire will give good results for a specific situation, complete photometric documentation on all the luminaires being considered is obviously needed. However, the way in which this information is presented to the user is just as important as the amount of data conveyed. National and international recommendations on road lighting vary considerably with regard to the way in which design parameters are quoted, and the photometric information should, ideally, be presented in the documentation in such a way as to make it universally applicable with the minimum of interpretation.

For conventional road-lighting luminaires two kinds of photometric documentation can be distinguished:

(a) Photometric specification of individual luminaires – data sheets
(b) Photometric performance of typical installations as calculated for different luminaire types – performance sheets.

Photometric data sheets

General data. The luminaire manufacturer often includes in his data sheet, in addition to his own luminaire type designation, a note on the luminaire's classification as defined by one or other of the systems described in the foregoing section. A note on the specific luminaire index (SLI) as defined by the CIE is particularly valuable as it facilitates the calculation of the glare control mark (G) for a given application.

Specifying the direction for which luminous intensity is a maximum, and the value of this maximum, furnishes the engineer with further valuable information concerning the general character of the luminaire. The luminous intensity can be specified as an absolute value (so many candelas) for the lamp type actually fitted, but in order to facilitate comparison between various luminaire types it is more usual to quote intensity as the value obtained for a specified nominal lamp flux, usually 1000 lumens, intensity then being specified as so many cd/1000 lm.

Also included on the data sheet is a note of the luminaire's upward and downward light output ratios. For the latter, it is usual to specify also how the flux is divided between the road and kerb sides of the lighting column.

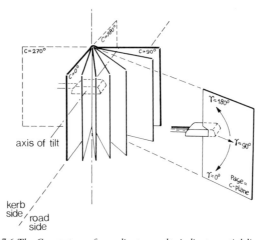

Figure 7.6 The C-γ system of coordinates used to indicate spatial directions.

In the interests of user convenience, the main bulk of the information contained in the data sheets is generally presented in graphical form using one of several standard systems for specifying spatial directions. The most preferred system for road-lighting luminaires, the so-called C-γ system, is illustrated in figure 7.6.

The graphs will normally contain information on each of the following: luminous intensity, illuminance, and road surface luminance.

Luminous intensity distribution. The luminous intensity distribution of a road-lighting luminaire is presented in two forms: as a polar light distribution diagram and as an isocandela diagram.

The polar diagram (figure 7.7) gives the intensity distribution for three vertical planes: the C-plane parallel to the road axis ($C = 0°$ and $C = 180°$); the C-plane at right angles to the road axis ($C = 90°$ and $C = 270°$); and the C-plane in which the maximum intensity lies.

108

A more detailed description of a luminaire's intensity distribution is contained in the isocandela diagram, which is a diagram showing the complete distribution of the luminaire in one hemisphere.

The isocandela diagram is constructed by projecting the theoretical surface surrounding the luminaire onto a plane (as in maps of the world). Lines are then drawn on the plane linking points of equal luminous intensity (isocandela contours).

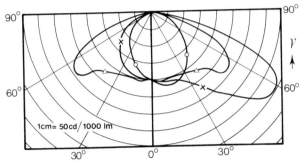

Figure 7.7 Example of a polar light distribution diagram, showing light distribution in plane parallel to the road axis (△), in plane at right angles to the road axis (○) and in plane in which I_{max} lies (×).

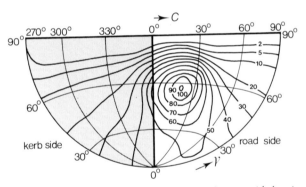

Figure 7.8 Example of an isocandela diagram in equal-area zenithal projection.

Several different types of projection are possible, that preferred for road-lighting applications being the so-called equal-area zenithal projection shown in figure 7.8. In this projection the plane is tangential to the sphere, perpendicular to the road surface, and at right angles to the road axis. Degrees along the 'equator' of the diagram give C-angles from the road axis, while degrees round the circumference give γ angles measured from the downward vertical.

The fact that equal areas on the sphere surrounding the luminaire are represented by equal areas on the projection (i.e. equal solid angles), means that the luminous flux in a given zone can be quickly and easily calculated, using the formula

$$\phi = \frac{2A}{r^2} \cdot I$$

where ϕ = luminous flux (in lm) in zone
       A = area of zone (in cm^2) on diagram
       r = radius (in cm) of the diagram
       I = average luminous intensity (in cd) in zone

A further useful feature of this projection is that the edges of a straight road appear on the diagram as straight lines. This means that the throw and spread of the luminaire can be easily determined from it, as indicated in figure 7.9. The isocandela diagram as illustrated in figure 7.8 is valid for an angle of tilt of zero degrees (axis of tilt is indicated in figure 7.6). Where this angle is anything other than zero, the grid of the diagram must be rotated, about the origin, through the appropriate angle relative to the contour lines to establish the new C-γ coordinates.

An alternative form of construction used for isocandela diagrams is that in which the isocandela contours are projected on the plane illuminated by the luminaire, viz. the road surface (figure 7.10). Such a diagram facilitates the classification of a luminaire according to the IES system of North America (Sec. 7.2.1). It can also be used for the construction of an isoluminance diagram for any arbitrary observer position (Sec. 13.2.2).

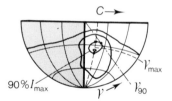

Figure 7.9 Throw (γ_{max}) and spread (γ_{90}) determined using an isocandela diagram of the equal-area zenithal projection type.

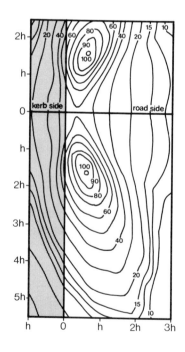

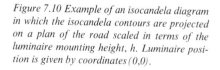

Figure 7.10 Example of an isocandela diagram in which the isocandela contours are projected on a plan of the road scaled in terms of the luminaire mounting height, h. Luminaire position is given by coordinates (0,0).

110

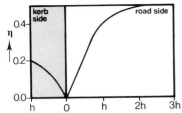

Figure 7.11 A typical utilisation factor diagram.

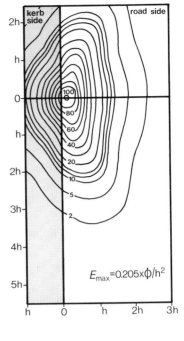

$$E_{max}=0.205\times\phi/h^2$$

Figure 7.12 Example of an isolux diagram showing the iso-illuminance contours at the road surface. Luminaire position is given by coordinates (0,0). The diagram is scaled in terms of the luminaire mounting height, h.

Illuminance data. The average illuminance given by the luminaire on a specified width of road, and the illuminance at any point on the surface covered by the luminaire can be quickly and easily calculated if the data sheet contains utilisation factor and isolux diagrams.

The utilisation factor of the luminaire, defined in road lighting as that fraction of the flux coming from the luminaire that actually reaches the road surface direct, is given in the utilisation factor diagram (figure 7.11) as a function of road width, which is specified in terms of luminaire mounting height to make the diagram valid for all mounting heights. (The calculation of average illuminance using this diagram is described in Sec. 13.2.1.)

The illuminance at a point on the road surface, and thus also the illuminance uniformity with a certain luminaire spacing, can be determined using the isolux diagram (figure 7.12). This diagram gives the iso-illuminance contours at the road surface, with the illuminances usually being expressed as a percentage of the maximum illuminance, the absolute value of which is given in terms of mounting height and lamp flux.

Luminance data. Average road-surface luminance and luminance uniformity can be calculated using the information contained in luminance yield factor and isoluminance diagrams, which are valid for specified observer positions.

Strictly speaking, the luminance yield factor is not a luminaire characteristic, but a characteristic of the installation. It is defined, for a single-sided lighting installation (figure 7.13), by

$$\eta_L = s\, L_{av}\, \frac{x}{Q_o \times \phi}$$

where η_L = the luminance yield factor of a longitudinal strip of the road
   s = the luminaire spacing (in m)
   L_{av} = the average road-surface luminance of the longitudinal strip (in cd/m^2)
   x = the width of the longitudinal strip (in m)
   Q_o = the average luminance coefficient of the road
   ϕ = the bare-lamp flux (in lm)

Since L_{av} is directly proportional to the reciprocal value of s, the product sL_{av} is constant for a given luminaire type.

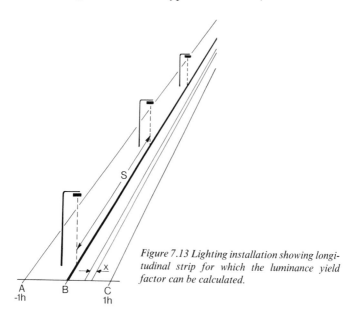

Figure 7.13 Lighting installation showing longitudinal strip for which the luminance yield factor can be calculated.

The luminance yield factor can therefore be considered not an installation characteristic but a general luminaire characteristic that indicates the effectiveness of the luminaire with regard to the creation of road-surface luminance.

The luminance yield factor diagram (figure 7.14) gives the luminance yield factor as a function of road width (this being in terms of luminaire mounting height) for a specified type of road surface and for three different positions of

112

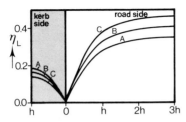

Figure 7.14 A luminance yield diagram in respect of three lateral positions of the observer:
A – observer 1 h to the left of the row of luminaires
B – observer in line with the row of luminaires
C – observer 1 h to the right of the row of luminaires

Figure 7.15 Example of an isoluminance diagram, with grid marked in terms of luminaire mounting height h. The position of the luminaire is given by coordinates (0,0) and the observer by coordinates (10 h, 0) – not shown

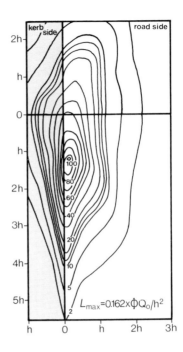

$$L_{max} = 0.162 \times \phi Q_o / h^2$$

the observer. The diagrams are available for each of the four standard road-surfaces with which all dry road surfaces can be represented as will be explained in Chapter 8. The calculation of average road-surface luminance using these diagrams is described in Sec. 13.2.2.

The luminance pattern produced by the luminaire on a specified type of road surface for a defined observer position is given in the isoluminance (or iso-cd/m²) diagram (figure 7.15). The diagram shows lines linking points of equal luminance (isoluminance contours) on a grid marked in terms of luminaire mounting height. Luminances are normally specified as a percentage of the maximum luminance. The absolute value of the maximum luminance is given in terms of the mounting height, luminous flux, and (See Chapter 8) the average luminance coefficient Q_o of the road surface. Diagrams are usually given for each of the four standard road surfaces mentioned above.

Performance sheets
Useful as the photometric data sheets are in calculating the various quantity and quality parameters of an installation, the job of design is made much easier if the lighting engineer is in possession of a set of precalculated performance sheets for the range of luminaires being considered.

Performance sheets may take many forms, but that reproduced in figure 7.16 will serve to illustrate the general features. The sheet is drawn up for a specific road cross-section, lighting arrangement, luminaire-lamp combination, and

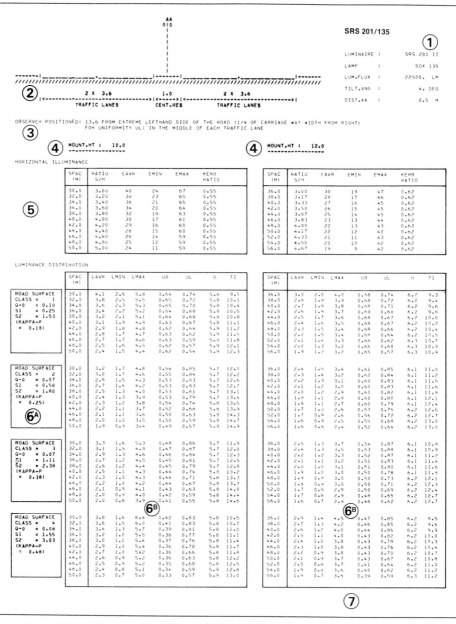

Figure 7.16 A Performance Sheet showing the general layout:
1 Type of luminaire, light source and general installation data
2 Cross section of the road showing lighting arrangement
3 Specification of observer position
4 The two mounting heights for which the sheet is valid
5 Illuminance values
6a Class of road surface
6b Luminance, uniformity and glare values
7 Code number for filing purposes

114

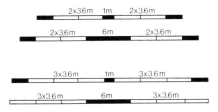

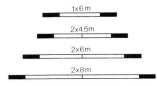

Figure 7.17 Suggested road cross-sections for which performance sheets could be drawn up.

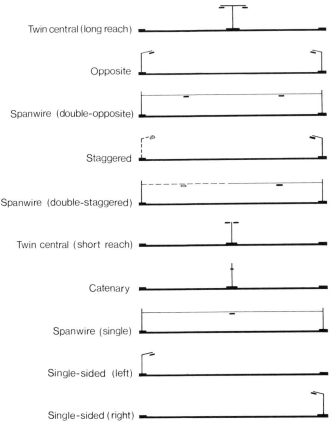

Figure 7.18 Suggested lighting arrangements for which performance sheets could be drawn up.

115

two luminaire mounting heights. The performance data is set out in tables: illuminance values above, luminance values and glare ratings below. The former (average illuminances and illuminance uniformities) are calculated by computer for a range of luminaire spacings. The same spacings have been used for calculating the luminances (average road-surface luminances and luminance uniformities) and glare ratings, but here the calculations have been repeated for each of the four standard road-surfaces mentioned earlier. A comprehensive library of such performance sheets, grouped according to, say, road type and covering a range of luminaire types, typical road cross-sections, and lighting arrangements (figures 7.17 and 7.18) offers two major benefits: one, the user has a conveniently packaged survey of the application possibilities of the luminaires covered; and two, he is provided with a frame-work of design solutions with which to work when seeking to satisfy specified quantity and quality requirements, and thereby a means of speeding up the entire design process. (More detailed information on performance sheets in connection with the design process of road lighting is given in Sec. 13.3.)

7.3 Floodlights

The principal difference in use between floodlights and the other luminaire types employed in road lighting is that the former are not aimed in standard directions with regard to the road axis. Instead, they are individually aimed at precalculated spots on the road surface, and this means that the photometric data supplied by the floodlight manufacturer must relate to the floodlight itself and not, as is the case with other road lighting luminaires, to a defined mounting position.

The photometric data of primary interest to the road lighting engineer are the peak intensity, beam spread, beam efficiency, and luminous intensity distributions of the unit concerned.

Peak intensity
The peak intensity of a floodlight occurs in the direction of the beam axis and is generally specified in candelas per 1000 lumens of lamp flux.

Beam spread
The beam spread of a floodlight, or in more common terms its beam width or beam angle, is the angle over which the luminous intensity drops to a stated percentage (usually 50 or 10 per cent) of its peak value. For a floodlight having a rotationally symmetrical luminous intensity distribution (that is a distribution that remains unchanged no matter what plane through the beam axis is considered) one figure for beam spread may be quoted; for example, 50° (viz. 25° to both sides of the beam axis). For an

asymmetrical distribution, as given by rectangular floodlights, two figures will be given; for example, 6°/24°, for the beam spreads in the two mutually perpendicular planes of symmetry (vertical and horizontal respectively). Sometimes the distribution in the vertical plane of such a floodlight is asymmetrical relative to the beam axis. Then two figures for the beam spread in this plane will be given; for example 5°–8°/24°, viz. 5° above and 8° below the beam axis and in the horizontal plane 12° left and 12° right from the beam axis.

The terms 'narrow-beam', 'medium-beam', and 'wide-beam' are frequently used to describe the beam spread of a floodlight in the plane of interest. An often used, but by no means generally accepted, definition of these terms is that based on the 50 per cent peak intensity beam-spread value:

	Beam spread at 50 % I_{peak}
narrow-beam	$\leqslant 20°$
medium-beam	$20°$ to $40°$
wide-beam	$\geqslant 40°$

Beam efficiency
The beam efficiency or light output ratio of a floodlight is defined as the ratio of the beam flux to the lamp flux. Three figures may be quoted: the total light output ratio, in which the total flux emitted by the floodlight is considered; and the light output ratios for the beam at 10 per cent and 50 per cent of peak intensity.

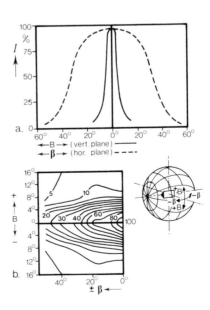

Figure 7.19 Floodlight luminous intensity distribution diagrams: (a) Diagram showing the relative luminous intensity of the unit in two planes mutually at right angles (b) Isocandela diagram showing the luminous intensity distribution at the surface of a sphere with the floodlight located at its centre, projected on a plane.

117

Luminous intensity distribution

The beam-spread characteristics of a floodlight serve as a useful guide to its selection, but for lighting design purposes a more detailed knowledge of the unit's intensity distribution is required. The photometric data sheet will therefore normally contain two luminous intensity distribution diagrams, one for each of the two planes of synmetry of the floodlight (figure 7.19a), and an isocandela diagram showing its intensity distribution at the surface of a sphere with the floodlight located at its centre projected on a plane (figure 7.19b).

In the case of floodlights, glare control is normally achieved by exercising special care in the positioning and aiming of individual units. In especially difficult situations, further control can be gained by using clip-on louvres to give the required degree of shielding.

Chapter 8

Road Surfaces

It may at first seem strange to find 'road surfaces' dealt with in this Part of the book, headed 'Equipment'. But it should be appreciated that the road surface in fact plays an essential role in determining the final lighting effect. The brightness of the road surface is an important quality determining factor with regard to both visual performance and visual comfort (Chapter 1), and it is determined by the light directed towards the surface from the lamps and luminaires of the installation and by the reflection properties of the road surface.

A theoretical approach to the study of the reflection properties of road surfaces has yet to lead to a workable system for describing these properties in quantitative terms. So far as road lighting is concerned, however, it is possible to work with a simplified description system. A classification system for all dry road surfaces, based on this description system, can be used to facilitate all dry-weather luminance calculations met with in road lighting. This description and classification system for dry weather conditions is described in detail in this chapter. A possible extension to the system to make it valid for wet weather conditions is also discussed.

8.1 Reflection Characteristics

There have been numerous attempts made to explain and characterise the reflection properties of various types of surface. Bouguer (1760), in a study involving rough surfaces, postulated that every surface could be considered as consisting of a great number of differently oriented particles or facets, each reflecting specularly.

Far more recent studies (Kebschull, 1968 and Vermeulen, 1975) have extended this earlier theory by suggesting that both specular and diffuse reflection takes place at these facets, the diffuse reflection being due to scattering on the facets.

Unfortunately, while this new theory serves to explain more fully the process underlying certain types of reflection, it has not, as yet, led to a technique for describing reflection in quantitative terms.

Another line of investigation (Sabey, 1971) involved the evaluation of surface texture. A surface can be described in terms of its macro and micro-texture (figure 8.1). Under dry conditions, both textures are instrumental in determining the way in which light incident upon the surface will be reflected. When wetted, however, the micro-texture floods and becomes specular, making reflection primarily macro-texture dependent. Useful as this new insight into the reflection process is in understanding reflection phenomena in general, it suffers from the practical drawback that micro-texture, unlike macro-texture, cannot be visually assessed or measured in a way suitable for practical lighting design (RRL, 1963).

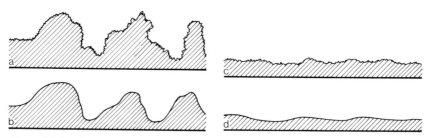

Figure 8.1 Road surface texture described in terms of the macro-micro structure: (a) rough-rough (b) rough-smooth (c) smooth-rough (d) smooth-smooth.

At present, the only way to completely specify the reflection characteristics of a road surface is by means of a set of so-called luminance coefficients. The luminance coefficient, q, is defined as the ratio of the luminance at an element on the surface to the illuminance at the same element, as given by a single light source. Thus

$$q = \frac{L}{E}$$

The luminance coefficient depends upon the nature of the road-surface material and upon the positions of the light source and the observer relative to the element under consideration, as defined by the four angles α, β, γ, δ in figure 8.2. Most road surfaces are almost completely isotropic, which means that the influence of angle δ can usually be neglected, particularly since for road widths up to as much as 25 metres and observation distances greater than 60 metres this angle is never greater than 20°. Another angle whose influence on q can in practice be neglected is α, the angle of observation. For the stretch of road important to a road user (60 m to 160 m ahead) and a standardised viewing height (1.5 m), α can range from 0.5° to 1.5°. Since, however, measurement has shown (Bergmans, 1938 and De Boer, *et al.*, 1952) that the α-dependency of q can be neglected over this range, it is usual for luminance coefficients to be determined with α held constant at 1°.

For the conditions of observation pertaining to car drivers, the luminance coefficient of a road surface can thus be said to be dependent upon only two angles, β and γ:

$$q = q\,(\beta, \gamma)$$

The complete reflection characteristics of a road surface could, therefore, be given in a table in which q values are specified for a number of β–γ combinations.

The calculation of luminance is made somewhat easier if the so-called reduced luminance coefficient (r) is used in place of q, where

$$r = q\cos^3\gamma$$

The luminance (L) at an element on the road given by a single source can then be written

$$L = qE = q\,\frac{I}{h^2}\cos^3\gamma = r\,\frac{I}{h^2}$$

where r = the reduced luminance coefficient of an element on the road surface as defined by β and γ (cd/m²/lux)

           I = the luminous intensity of the luminaire in the direction of the point (cd)

           h = the mounting height of the luminaire (m)

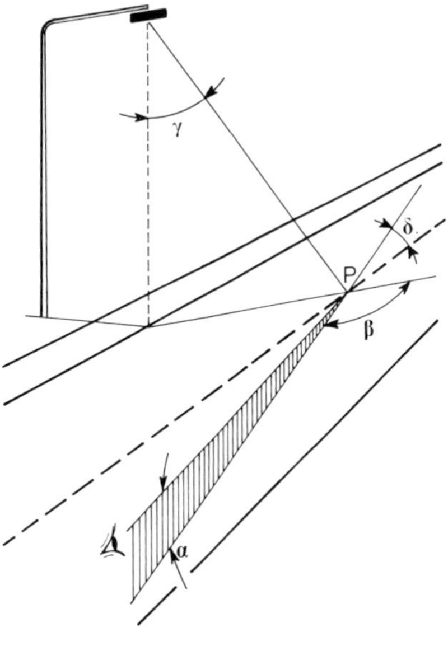

Figure 8.2 Angles upon which the luminance coefficient of a road surface is dependent:
α = angle of observation (from the horizontal)
β = angle between plane of light incidence and plane of observation
γ = angle of light incidence
δ = angle between plane of observation and road axis

ROADSURFACE STANDARD R2

BETA DEG.	0.0 / 7.50	0.25 / 8.00	0.50 / 8.50	0.75 / 9.00	1.00 / 9.50	1.25 / 10.00	1.50 / 10.50	1.75 / 11.00	2.00 / 11.50	2.50 / 12.00	3.00	3.50	4.00	4.50	5.00	5.50	6.00	6.50	7.00
0	5571 (957)	5871 (909)	5871 (828)	5414 (785)	4785 (742)	4328 (700)	3871 (671)	3557 (628)	3242 (600)	2785 (585)	2285	2085	1885	1685	1514	1371	1242	1114	1014
2	5571 (542)	5871 (471)	5871 (400)	5414 (357)	4785 (328)	4328 (300)	3871 (257)	3400 (228)	3085 (200)	2714 (185)	2214	1871	1614	1357	1157	985	828	714	614
5	5571 (171)	5871 (142)	5871 (128)	5414 (100)	4785 (85)	4171 (71)	3714 (71)	3242 (57)	2785 (57)	2085 (57)	1642	1242	957	714	542	414	314	242	200
10	5571 (57)	5871 (42)	5871 (42)	5257 (42)	4642 (28)	3871 (28)	3242 (28)	2785 (28)	2171 (28)	1571 (14)	957	585	385	285	200	157	114	85	71
15	5571 (42)	5871 (28)	5757 (28)	5100 (28)	4171 (28)	3400 (14)	2557 (14)	2171 (14)	1671 (14)	1057 (14)	614	357	214	171	114	85	71	57	42
20	5571 (28)	5871 (28)	5757 (28)	4942 (14)	4157 (14)	2942 (14)	2171 (14)	1771 (14)	1357	828	471	257	171	128	85	71	57	42	42
25	5571 (28)	5871 (28)	5485 (28)	4642 (14)	3714	2628	2014	1514	1142	685	371	214	142	100	71	57	57	42	28
30	5571	5871	5414	4328	3400	2171	1700	1300	957	571	300	185	128	100	71	57	42	42	28
35	5571	5872	5285	4014	3085	1857	1542	1114	871	500	257	171	128	85	71	57	42		
40	5571	5871	4942	3714	2785	1700	1328	957	742	428	242	157	114	85	71	57			
45	5571	5414	4642	3400	2471	1542	1142	871	642	385	228	157	114	85	71				
60	5571	5257	4328	3085	2171	1428	1085	742	571	342	228	157	114	85	57				
75	5571	5100	4014	2942	2171	1471	1085	771	585	371	242	157	114	85	71				
90	5571	5100	4014	2942	2171	1514	1142	828	642	400	242	157	128	85	71				
105	5571	4942	3871	2942	2171	1542	1200	900	700	428	257	171	142	100	85				
120	5571	4942	3871	2942	2171	1542	1242	957	742	471	300	200	157	114	100				
135	5571	4942	3871	2942	2014	1628	1271	985	771	500	314	214	171	142	128				
150	5571	4785	3714	2942	2014	1628	1300	1014	800	542	342	242	185	171	142				
165	5571	4785	3714	2942	2014	1700	1328	1042	814	571	371	257	214	185	142				
180	5571	4785	3714	2942	2014	1700	1357	1057	828	585	385	300	242	200	157				

Column headers above are labelled TAN (GAMMA).

ALL VALUES ARE MULTIPLIED BY 10000

Figure 8.3 Example of a table of reduced luminance coefficients for standard road surface R2.

122

An example of a reduced luminance coefficient table (also known as a reflection table, or *r*-table) is given in figure 8.3. It will be noticed that the coordinates of each *r* value are given not as two angles, β and γ, but as angle β and $\tan\gamma$ (BETA DEG. and TAN (GAMMA) in the language of the computer). The combination of β and $\tan\gamma$ at which *r*-values are given in figure 8.3 is in accordance with a recommendation appearing in CIE Publication No. 30 – (CIE, 1976a).

Using a table such as this (drawn up for a particular road surface) in conjunction with the formula given above, it is a simple matter to calculate the luminance produced by a single luminaire at any point on the surface; repeating this for adjacent luminaires and summing the results will then give the total luminance at the point.

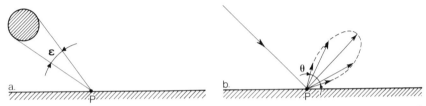

Figure 8.4 The influence that source luminance has on the magnitude of the luminance coefficient can be safely neglected so long as the solid angle ε subtended by the source at the point of measurement (a) is small compared with the solid angle θ within which a parallel beam of light from the source would be reflected by the surface at this point (b).

The reflection table, or table of luminance coefficients, thus fully defines the reflection characteristics of a road surface and so makes luminance calculations possible. There is, however, one important restriction to the use of luminance coefficients in this way that should be borne in mind (Bergmans, 1938); namely, that for specular or near-specular road surfaces the luminance coefficient ceases to be a characteristic of the surface alone, and becomes dependent upon the luminance and size of the source providing the illumination. As the road surface becomes more diffuse, the influence of the luminance of the light source (and hence, at constant luminous intensity, its physical size) decreases, but only disappears completely for a perfect diffuser. In practice, however, this influence that the luminance of the source has on the magnitude of the luminance coefficient can be safely neglected when the specularity of the road surface is such that the solid angle (ε) subtended by the source at the point of measurement (figure 8.4a) is small compared with the solid angle (θ) within which a parallel beam of light from the source would be reflected by the surface (figure 8.4b) (De Boer *et al.*, 1967a) – a condition that is in practice fulfilled for all types of luminaire so long as the road surface is not inundated.

8.1.1 Dry Surfaces

The *r*-values needed in order to draw up a reflection table can only be accurately determined by performing lengthy and tedious laboratory measurements on a representative sample of the road surface (see Sec. 14.1.2). There is, however, an alternative to using exact, tailor-made tables. It has been shown that the reflection characteristics of most dry road surfaces can be reliably represented using just three parameters, the so-called description parameters. It has also been shown that just one of these same parameters can be used to classify any dry surface as belonging to one of four groups or classes. Each of these four classes has been assigned a standard reflection table. Thus, all that the lighting engineer need in fact do as a preliminary to making his luminance calculations, is examine his road surface, decide to which of the four classes it belongs, and base his design on the standard reflection table of the relevant class.

Description parameters

Since the publication of the so-called Q zero – Kappa p system for describing the reflection characteristics of road surfaces (Westermann, 1963; De Boer and Westermann, 1964a and b; and De Boer and Vermeulen, 1967b), a number of workers have come up with new or revised description systems (Roch and Smiatek, 1972; Range, 1972; Erbay, 1973 and 1974b; Massart, 1973; and Burghout, 1977a).

The CIE (1976a) adopted a slightly modified version of Erbay's system, which in turn can be said to be a revision of the Q zero – Kappa p system. Burghout (1977b), in a recent study in which he evaluated the inaccuracies accompanying various systems, has shown that this CIE system forms a sound basis for a classification system for dry road surfaces. Since classification is in fact the final goal of a description system, this internationally accepted system of the CIE will be dealt with here.

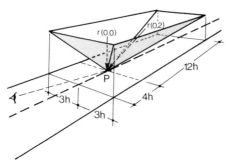

Figure 8.5 The solid angle Ω_o in the definition of Q_o. This can be defined as the angle subtended at point P by a 'ceiling' containing all those luminaires that contribute to the Q_o value at this point on the road surface. The ceiling extends a distance 3h to the right and left of the point, 4h towards the observer and 12h away from him. Luminaires outside this area normally contribute very little to the luminance at the point and can therefore be safely neglected. Also indicated in this figure are the directions of those reduced luminance coefficients that play a role in determining the specular factors S1 and S2.

The choice of parameters used to describe or summarise the reflection characteristics of a road surface – or, looked at another way, to characterise the
contents of a reflection table – is based on the consideration that most
surfaces can be reasonably well described in terms of two basic qualities: their
lightness (or degree of greyness from white to black), and their specularity (or
shininess). The parameters used in the CIE system to express these qualities
numerically are

Q_0 (the average luminance coefficient), for the lightness and
$S1$ and $S2$ (specular factors), for the specularity

where $\quad Q_0 = \dfrac{1}{\Omega_0} \displaystyle\int_0^{\Omega_0} q \, d\Omega$

$S1 = \dfrac{r(0,2)}{r(0,0)}$

$S2 = \dfrac{Q_0}{r(0,0)}$

in which q = luminance coefficient (dependent upon β and γ)
Ω_0 = solid angle, measured from a point on the surface, containing
all those directions from which light is incident that are taken
into account in the averaging process (see figure 8.5 for
clarification)
$r(0,2)$ = value of reduced luminance coefficient with $\beta = 0°$ and $\tan \gamma$
$= 2$ (see figure 8.5)
$r(0,0)$ = value of reduced luminance coefficient with $\beta = 0°$ and $\tan \gamma$
$= 0$ (see figure 8.5)

The average luminance coefficient is thus defined as the solid angle weighted
average of the luminance coefficients. This ensures that a balanced contribution of q values is obtained, viz. the large q values, corresponding to large γ
angles (fig. 8.2) do not have an over-dominating influence on the value of Q_0.
In practice, Q_0 is often calculated as a function of the reduced luminance
coefficient using r-tables in which r is defined for a range of β-$\tan\gamma$ combinations, the equation linking Q_0 and r being derived as

$$Q_0 = \frac{1}{\Omega_0} \int_0^{\Omega_0} q \, d\Omega = \frac{1}{\Omega_0} \int_0^{\Omega_0} r(\beta, \tan\gamma) \tan\gamma \, d(\tan\gamma) \, d\beta$$

The limits of the rectangular solid angle Ω_0, expressed in β and $\tan\gamma$ limits,
can, for ease of working, be derived from figure 8.6. For these limits table 8.1
gives weighting factors as given by Sørensen (1974) which can be used to
calculate the Q_0 value from reflection tables.

125

Table 8.1 Weighting factors used when calculating the luminance coefficient Q_o from reflection tables

BETA → TANGAMMA	0	2	5	10	15	20	25	30	35	40	45	60	75	90	105	120	135	150	165	180
.00	8	8	32	22	40	20	40	30	40	25	45	120	60	120	60	120	60	120	60	60
.25	32	32	128	88	160	80	160	80	160	100	180	480	240	480	240	480	240	480	240	240
.50	16	16	64	44	80	40	80	40	80	50	90	240	120	240	120	240	120	240	120	120
.75	32	32	128	88	160	80	160	80	160	100	180	480	240	480	240	480	240	480	240	240
1.00	16	16	64	44	80	40	80	40	80	50	90	240	120	240	120	240	120	240	120	120
1.25	32	32	128	88	160	80	160	80	160	100	180	480	240	480	240	480	240	480	240	240
1.50	16	16	64	44	80	40	80	40	80	50	90	240	120	240	120	240	120	240	120	120
1.75	32	32	128	88	160	80	160	80	160	100	180	480	240	480	240	480	240	480	240	240
2.00	24	24	96	66	120	60	120	60	120	75	135	360	180	360	180	360	180	360	180	180
2.50	64	64	256	176	320	160	320	160	320	200	360	960	480	960	480	960	480	960	480	480
3.00	32	32	128	88	160	80	160	80	160	100	120	510	222	240	180	480	240	480	240	240
3.50	64	64	256	176	320	160	320	160	320	200	120	270	33	0	−30	690	480	960	480	480
4.00	32	32	128	88	160	80	160	80	160	100	60	0	0	0	−75	−30	255	510	222	120
4.50	64	64	256	176	320	160	320	160	320	155	75	0	0	0	0	0	135	372	33	0
5.00	32	32	128	88	160	80	160	105	125	5	−15	0	0	0	0	0	0	33	0	0
5.50	64	64	256	176	320	160	320	170	90	0	0	0	0	0	0	0	0	0	0	0
6.00	32	32	128	88	160	80	160	40	0	0	0	0	0	0	0	0	0	0	0	0
6.50	64	64	256	176	320	160	275	35	0	0	0	0	0	0	0	0	0	0	0	0
7.00	32	32	128	88	160	105	80	−25	0	0	0	0	0	0	0	0	0	0	0	0
7.50	64	64	256	176	320	170	90	0	0	0	0	0	0	0	0	0	0	0	0	0
8.00	32	32	128	88	160	85	45	0	0	0	0	0	0	0	0	0	0	0	0	0
8.50	64	64	256	176	320	80	0	0	0	0	0	0	0	0	0	0	0	0	0	0
9.00	32	32	128	88	160	40	0	0	0	0	0	0	0	0	0	0	0	0	0	0
9.50	64	64	256	176	320	80	0	0	0	0	0	0	0	0	0	0	0	0	0	0
10.00	32	32	128	88	115	−5	0	0	0	0	0	0	0	0	0	0	0	0	0	0
10.50	64	64	256	176	230	−10	0	0	0	0	0	0	0	0	0	0	0	0	0	0
11.00	32	32	128	113	80	−25	0	0	0	0	0	0	0	0	0	0	0	0	0	0
11.50	64	64	256	186	90	0	0	0	0	0	0	0	0	0	0	0	0	0	0	0
12.00	16	16	64	69	45	0	0	0	0	0	0	0	0	0	0	0	0	0	0	0

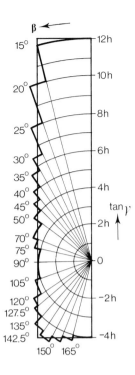

Figure 8.6 Integration boundaries in terms of tan γ and β as projected on the road, for the calculation of the average luminance coefficient Q_o.

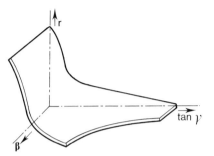

Figure 8.7 Three-dimensional plot or 'r-model' representing the information contained in the r-table of figure 8.3.

Figure 8.8 r-models corresponding to two extremes of dry road surface (a) a very glossy surface (b) a very diffuse surface.

Figure 8.9 r-models corresponding to two surfaces of equal specularity (same shape) but differing in lightness, model (c) being much lighter than model (d).

127

That it is indeed possible to summarise the reflection characteristics of a dry road surface by means of the Q_o, $S1$, and $S2$ parameters defined above, can be seen by studying the three-dimensional plot shown in figure 8.7. The β and $\tan\gamma$ axes of co-ordinates of the plot are the same as those used in the r-table of figure 8.3, while the vertical axis has been used to show the r-value of each point taken from the same table.

The 'r-model' so obtained thus represents the information contained in the r-table, and thus summarises the reflection characteristics of the road surface defined by that table. Of course, other types of dry road surface will yield other r-models. This is illustrated in figure 8.8, where r-models for two contrasting surfaces are shown: a very glossy surface (model a) and a very diffuse surface (model b).

The difference in glossiness, or specularity, between one surface and another is depicted in the models as a difference in shape: the more specular the road surface, the closer will the shape of its model resemble that of model a (the larger the r-values at large γ angles, i.e. large $\tan\gamma$ values, the more specular the surface). The two r-models shown in the figure in fact correspond closely to the extremes of dry road surfaces (most glossy and most diffuse occurring dry surface).

Since the difference in shape for these extremes of specularity is not very marked, it should be possible to indicate the shape corresponding to a given dry road surface by a single factor.

The factor needed is actually a ratio, the ratio of the r-value of a point on the model to the r-value at the origin; the greater this ratio, the closer will the r-model concerned approach the limit represented by model a in figure 8.8. A convenient ratio to take, is that defined by

$$\frac{r(\beta = 0°, \tan\gamma = 2)}{r(\beta = 0°, \tan\gamma = 0)}$$

which is the definition of the specular factor ($S1$) given earlier. The value of $S1$ is thus a comparative indication of the degree of specularity of the road surface to which it belongs: the larger the value of $S1$, the more specular the surface.

Just as the comparative shape of the r-model is an indication of the specularity of the road surface, so is its height above the β-$\tan\gamma$ plane (datum) an indication of that surface's lightness. This is illustrated in figure 8.9 in which two r-models are shown for two surfaces that are equal in specularity (i.e. they have the same shape) but which differ in lightness, the surface belonging to model c being the lighter of the two. Now the height of the model above datum is directly proportional to the volume within it, and this is given by

$$\int_0^{\Omega_0}\int r \cdot d(\tan\gamma)d\beta$$

3 Motorway in Belgium with double twin-central arrangement of low-pressure sodium luminaires. (See Sec. 15.1.1)

4 Catenary lighting in the Netherlands employing twin-lamp low-pressure sodium luminaires suspended 15 metres above the ground and spaced 22.5 metres apart. Pole spacing: 90 metres. Total width of the road: 74 metres. (See Sec. 15.1.

where Ω_0 represents the integration boundaries of β and γ. Comparing this formula with the definition of Q_0 given earlier,

$$\text{i.e. } Q_0 = \frac{1}{\Omega_0} \int_0^{\Omega_0} \int r \tan\gamma \, d(\tan\gamma) \, d\beta$$

it can be seen that Q_0 is in fact indicative of the above volume, and is thus a measure of the lightness of the road surface concerned.

It will be clear from what has been said above that the $S1$–Q_0 description system has certain inherent inaccuracies (the practical consequences of which will be evaluated below under the heading Classification). It was to make the system more precise, therefore, that the CIE introduced the extra shape-defining parameter, $S2$, defined earlier (in fact $S2$ equals $10^{\text{Kappa}-p}$, where Kappa-p is the specular factor used in the original De Boer/Westermann description system). However, since $S2$ and $S1$ are closely correlated (figure 8.10), the inclusion of the former does not serve to give any important additional information.

Parameters less closely related to $S1$, and therefore potentially more useful as extra parameters, have in fact been introduced by Burghout (1977a). But the same worker (Burghout, 1977b) has also shown that $S1$ on its own forms a sound basis for a classification system involving dry road surfaces. There therefore seems little point in using an additional parameter.

The r-model introduced here is just one of various types of presentation that can be used to explain the $S1$–Q_0 description system. When explaining the original Q_0-Kappa p system, for example, De Boer and Westermann used the reflection indicatrix shown in figure 8.11. Such a reflection indicatrix is also a graphical representation of a reflection table (but in terms of q rather than r); it can therefore be used in much the same way as the r-model to explain the $S1$–Q_0 system.

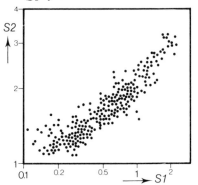

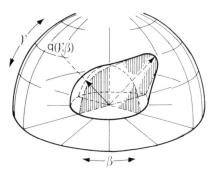

Figure 8.10 Correlation between the specular factors S1 and S2. Each point gives the S1 and S2 values of a measured sample of road surface.

Figure 8.11 Reflection or luminance coefficient indicatrix.

Classification

It has been shown above that the reflection properties of a road surface can be described using just three parameters: the specular factors $S1$ and $S2$, together with the average luminance coefficient, Q_0. (It has also been shown that the specular factor $S2$ does not contribute much to the accuracy of the description.) The only problem remaining is how to obtain the reflection table corresponding to these parameters, so that luminance calculations can be made with the desired degree of accuracy.

One possibility is to select from a large number of reflection tables, each drawn up using measured road surfaces, a table having specular factors corresponding as closely as possible to the $S1$ and $S2$ factors of the actual road surface considered. The r-values of the table selected must be adjusted to the actual Q_0 value of the road surface in question by multiplying each and every one by the ratio $Q_0 : Q_{0 \, tab}$ where Q_0 is the actual Q_0 value and $Q_{0 \, tab}$ the Q_0 value of the table selected.) A large set of reflection tables which can be used for this purpose is given by the University of Berlin (Erbay, 1974a) and by Sørensen (1975).

Comparing specular factors in this way is time consuming and suffers from the added disadvantage that different sets of tables will inevitably lead to different 'end tables' finally being selected for one and the same surface. This makes intercomparisons impossible unless the same tables are used by everyone for selection purposes.

A method that does not give the above-mentioned problems, and which can therefore be recommended, is that based on the classification of road surfaces into groups or classes. Each type of dry road surface is placed in one of four classes according to its $S1$ value. By accepting a certain inaccuracy it is possible to characterise each road surface in one of four classes by only one reflection table typical for that class. In other words, each class can be assigned a standard reflection table with which luminance calculations for surfaces falling in that class can be made.

The most suitable class limits and standard tables for each class depend to a certain extent on the type of road surfaces likely to be encountered in practice. The CIE specifies two sets of standard reflection tables, the so-called $R1$ to $R4$ and $N1$ to $N4$ tables. The corresponding class limits and the values of the description parameters for these tables are given in table 8.2. The complete R-tables and N-tables are to be found in Appendix A.

As can be seen from table 8.2, the N-standards, based on road surfaces found in Scandinavia where artificial surface brighteners are often employed to give very diffuse surfaces, have a class 1 with very small $S1$ values and a class 3 covering a wide $S1$ range (covering almost the entire class 3 and part of class 2 in the R-standards). It is recommended, therefore, that the R-classification

Table 8.2 Road surface class limits of the R and N-standards and the S1 values of the various standards, with an indication of the average Q_o value for each class

Class	R-standards		N-standards		R and N standards
	S1-limits	S1 of standards	S1-limits	S1 of standards	'Average' Q_o value
1	$S1 < 0.42$	0.25	$S1 < 0.28$	0.18	0.10
2	$0.42 \leqslant S1 < 0.85$	0.58	$0.28 \leqslant S1 < 0.60$	0.41	0.07
3	$0.85 \leqslant S1 < 1.35$	1.11	$0.60 \leqslant S1 < 1.30$	0.88	0.07
4	$1.35 \leqslant S1$	1.55	$1.30 \leqslant S1$	1.61	0.08

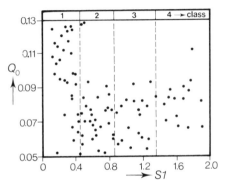

Figure 8.12 Q_o–S1 diagram (average luminance coefficient – specular factor) with indication of the four R-classes of road surface, together with a number of Q_o–S1 values of measured road surfaces.

system be used in countries where the use of artificial brighteners is the exception rather than the rule.

Again, the reflection tables must be rescaled according to the actual average luminance coefficient of the road surface for which the luminance calculations are to be made. However, since the standard reflection tables (see Appendix A) are valid for a Q_o value of 1.0, this merely involves multiplying through by the measured Q_o value. Where this is not known, the 'average' class value recommended by the CIE may be used (table 8.2 last column), it being possible to specify such an average by virtue of the correlation (although weak) existing between road surface class and Q_o, as shown in figure 8.12.

System inaccuracies

Some idea of the inaccuracies involved in the R-classification system can be obtained from a study of figure 8.13 in which the overall and longitudinal

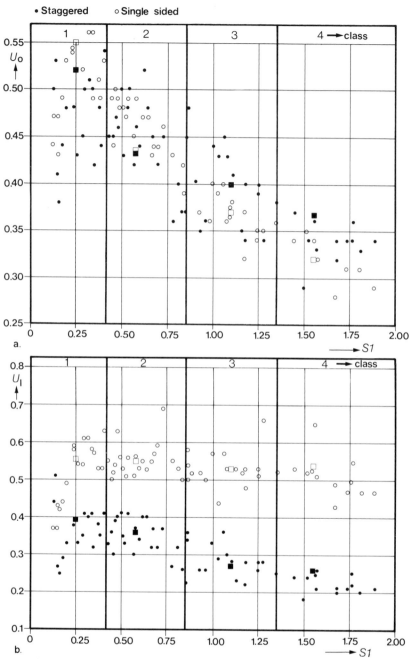

Figure 8.13 Calculated luminance uniformities: (a) overall and (b) longitudinal, for some sixty different road surfaces, each characterised by its S1 value. The results are valid for tight-controlled luminaires employed in single-sided (○) and staggered (●) lighting arrangements at a mounting height of 10 m and a spacing of 45 m on roads 10.8 m wide. The symbols □ and ■ indicate results obtained using standard R-tables for the single-sided and staggered lighting arrangements respectively.

132

luminance uniformities are given as calculated using a given luminaire in single-sided and staggered arrangements for some sixty different road surfaces spread over the four R-classes. The results for these different road surfaces are shown together with the results as would have been obtained with each of the four R-standards.

A more detailed insight into the inaccuracies accompanying the use of this classification system was gained as follows. From a combination of 44 widely differing light distributions and 113 different dry road surfaces no less than 4972 pairs of L_{av}, of U_o and of U_l were calculated, one value of each pair being obtained using the r-table corresponding to the road surface in question and the other using the R-standard for that surface. All calculations were made assuming a typical single-sided installation.

The difference between the two values in each pair was then expressed as a percentage of the r-table value and the results averaged for each of the three parameters to give a representative indication of the inaccuracy of the R-classification (or R-standard) system viz.

$$\text{parameter error} = \sum_{l=1}^{44} \sum_{r=1}^{113} \frac{|P_{r,l} - PS_{r,l}|}{P_{r,l}}$$

where $P_{r,l}$ = value of parameter considered (L_{av}, U_o or U_l) for road surface r and light distribution l

$PS_{r,l}$ = value of same parameter with same light distribution but for standard surface to which r belongs

The percentage inaccuracies for the four R classes are summarised in table 8.3.

Burghout (1979), in a very recent study, suggests that the inaccuracies do not dramatically increase if road surfaces belonging to classes R_2, R_3 and R_4 are all represented by one and the same standard table.

Table 8.3 Average quality parameter inaccuracies given by the R-classification system for the four standard road-surface classes

Standard road-surface classes	Error (%) in quality parameter		
	L_{av}	U_o	U_l
R1	5.4	8.6	10.6
R2	6.0	6.9	8.3
R3	5.7	7.4	9.2
R4	5.4	8.4	9.6

8.1.2 Wet Surfaces

In practice, of course, all road-surface conditions between dry and inundated will be met with. It has already been pointed out that for the inundated surface one can no longer talk about the luminance of the surface itself, but only about the luminances of the light sources mirrored in that surface. For wet but not inundated surfaces, on the other hand, the luminance coefficient can still be meaningfully defined, and this means that for these conditions valid luminance calculations can indeed be made using reflection table values. The problem is, however, that no simple method for arriving at the relevant wet-surface reflection table as yet exists, although investigations in this direction are being carried out, especially in Denmark. Some preliminary results will be discussed here.

The first requirement when designing for wet weather conditions is, of course, a definition of what exactly constitutes a wet surface. The CIE proposes for this purpose the condition of a sample of the road surface found 30 minutes after having been uniformly sprayed with water at a rate equivalent to 5 mm of rain per hour in a draught-free room at a temperature of 25 °C and with a relative humidity of 50 per cent (CIE, 1976a). This condition of the sample the CIE calls the standard wet condition. It is chosen on the basis of measurements performed on samples from Danish roads, from which it appeared that the specular factor $S1$ of a sample subjected to local night-time weather conditions was larger than the $S1$ value of the same sample in the standard wet condition for only 10 per cent of the time (Frederiksen and Gudum, 1972). It should be realised, therefore, that the practical meaning of the definition is dependent upon local climatological conditions, upon locally employed types of road surface, and upon traffic load conditions (roads with a high traffic density dry more quickly than do roads with a low density).

Frederiksen and Sørensen (1976) tested the accuracy of the $S1$, $S1$–$S2$, Q_0–$S1$, and Q_0–$S1$–$S2$ description systems on wet (standard wet condition) surfaces. They found the three-parameter system to be the most accurate, but because it is rather complicated to use, opted for the next most accurate Q_0–$S1$ system as the basis for their wet-surface classification system. These tests made allowance for the fact that Q_0 for wet surfaces loses its identity as a pure scaling factor and takes on the function of a combined scaling and specular factor. (It is, for example, possible to have two wet surfaces with the same $S1$ value but nevertheless with differently-shaped r-models, the difference in shape being accompanied by a difference in Q_0 value.)

Using these parameters, and working with samples of road surface types common to Scandinavia these workers established four road-surface classes valid for the standard wet condition. The classes were so chosen as to minimise errors in the values of certain important photometric quality parameters obtained when using the classification system.

The class limits of this system in terms of $S1$ and Q_o are given in figure 8.14. By using a so-called corrected specular factor, defined in terms of Q_o and $S1$ such that it has a constant value along any of the inclined lines of figure 8.14 (the borderlines of the classes), a one-parameter classification system is obtained.

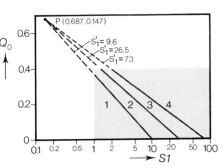

Figure 8.14 Class limits in the Danish wet-surface classification system proposed by Frederiksen and Sørensen. The shaded area is applicable to wet road surfaces and the inclined lines running through represent constant values of the corrected specular factor S1' which can be used to determine the class of such a surface.

This corrected specular factor, $S1'$, is chosen such that its absolute value corresponds to the $S1$ value of a class borderline at the point where it crosses the $S1$ axis (in the region where $S1$ is greater than 1). Thus

for $S1 > 1$

$$\log (S1'/0.147) = \frac{\log (S1/0.147)}{1 - Q_o/0.687}$$

and for $S1 \leqslant 1$

$$S1' = S1$$

where the $S1$ and Q_o values are those for the wet condition.

The class limits can be defined using the corrected specular factor, as shown in table 8.4.

Table 8.4 Wet road surface class limits in terms of the corrected specular factor S1'

Class	S1' limits
W1	$S1' < 9.6$
W2	$9.6 \leqslant S1' < 26.5$
W3	$26.5 \leqslant S1' < 73$
W4	$73 \leqslant S1'$

Frederiksen and Sørensen give a standard reflection table for each of the four wet road surface classes. This makes it possible to base luminance calculations for each wet road surface on the reflection table typical of the class to which the surface belongs, as decided by its $S1'$ value. These four standard reflection tables, $W1$ to $W4$, are given in Appendix B for a Q_o value (wet) equal to 1. As with dry surfaces, the tables must be rescaled to make them valid for the actual wet-surface Q_o value prevailing. (The corresponding dry surface Q_o value is also given in the four standard W-tables, Frederiksen and Sørensen having shown that the actual dry surface Q_o value can also serve as a scaling factor for the wet surface.)

The proposal of Frederiksen and Sørensen for a complete and unified classification system for both dry and wet surfaces referred to above is limited in practice by the fact that it is based on a rather arbitrarily defined standard wet condition (for samples, not for roads) and on road surface types typical for Scandinavia. At present it is the only in-depth material available on wet road surfaces. It is, as such, also described in CIE report 'Road lighting for wet conditions' (CIE, 1980). Before further progress can be made towards developing a general and perhaps more practicable system, further investigations into wet-weather road surface conditions are clearly needed.

8.2 Acquisition of Reflection Data

Before examining the ways in which the reflection properties of a particular surface may be determined in order to indicate the class to which the surface belongs, it is first necessary to consider the question of surface wear and the effect this will have on the results obtained.

Most surfaces will show a change in their reflection properties with wear, the change normally being greatest during the first period of use. The actual rate of wear for a given surface is virtually impossible to predict – it depends so much on local climatological conditions and type and density of traffic load – but some time after a surface is layed down, a point is reached when it can be said to have stabilised as far as any further change in its reflection properties is concerned. This point is usually reached after the first few months of use. This is the surface condition that should, of course, be taken for lighting design purposes; any seasonal variations in the surface's reflection properties that may occur after this initial 'running in' period are usually of second-order importance.

The change in reflection properties will not be constant over the whole road surface, and the difference is likely to be greatest across the road due to differences in wheel loading from one lane to another. Ideally, therefore, the reflection properties should be determined for a number of different locations on the road in order that its 'average' reflection properties may be defined. There are various ways of determining the reflection properties of a given road surface: laboratory measurements, in-situ measurements, and by a comparison of road-surface materials.

8.2.1 Laboratory Measurements

A complete reflection table can be drawn up based on measurements performed on a small sample of the surface concerned. (The measurement technique is described in Chapter 14.) This method is both complicated and time consuming: the sample must be cut and then transported to a suitably equipped lighting laboratory. Furthermore, care must be taken during this phase of the operation that the properties of the sample do not change so much as to make the subsequent measurements performed on it invalid, e.g. because of cracking or drying out. The sample (or samples) should be taken from a part of the road that is representative of the whole.

8.2.2 In-situ Measurements

In-situ measurements can be made with a reflectometer to determine the surface's $S1$ and Q_0 values ($S2$ as well, if needed). (A description of reflectometers will be found in Chapter 14.) The advantage of this method is that measurements can easily be carried out at different locations on the road, so allowing the 'average' reflection properties of the surface to be determined with a fair degree of confidence. The method also affords an insight into the way reflectance may vary from one spot to another.

8.2.3 Comparison of Materials

A rough idea of the reflection properties of a surface can be gained by studying the materials of which the surface is composed. CIE Publication No. 30 (CIE, 1976a) gives a table in which examples of road-surface materials are listed according to the probable dry road-surface class resulting from their use (table 8.5). Table 10.1 in Chapter 10 (Bad-Weather Lighting) gives an idea of the relationship between the composition of a road surface and its probable wet-surface class.

When virtually identical materials are used in the composition of similarly constructed road surfaces throughout a given area, it can be expected that roughly similar reflection properties will be obtained for all roads in that area. It follows, therefore, that once the class and Q_0 value of just some of the roads in the area has been determined by measurement, the same class and Q_0 value can be used for the roads remaining.

Table 8.5 Examples of road surface materials which possibly lead to a specific dry road surface class (source: CIE Publication 30)

Class	Materials
I	– Asphaltic type road surface containing either at least 15 per cent artificial brightener or at least 30 per cent very bright anorthosites. – Surface dressings containing chippings that cover over 80 per cent of the road surface, where the chippings mainly consist of either artificial brighteners or are 100 per cent very bright anorthosites. – Concrete road surfaces.
II	– Surface dressings having a harsh texture and containing normal aggregates. – Asphaltic surfaces containing 10 to 15 per cent artificial brighteners. – Coarse and harsh asphaltic concrete, rich in gravel (> 60 %) of sizes up to or greater than 10 mm. – New condition mastic asphalt.
III	– Asphaltic concrete (cold asphalt, mastic asphalt) having gravel sizes up to 10 mm, but of a harsh texture (similar to sand paper). – Polished, coarse textured, surface dressings.
IV	– Mastic asphalt after some months of use. – Road surfaces having a rather smooth or polished texture.

Part 3

DESIGN

Chapter 9

Recommendations

The investigative work carried out in connection with each of the main criteria relevant to the lighting of roads and tunnels, viz. level, uniformity and glare, and the quantitative influence that these criteria have on the visual performance and visual comfort of a road user have been described in detail in Part One of this book.

The present chapter goes a step further and summarises the most important findings to emerge from this research as seen in the light of present-day knowledge and experience. The chapter at the same time presents the reader with a practical survey of those lighting criteria that must be considered when actually designing a lighting installation.

Many of the findings included in the survey are seen reflected in the latest international recommendations for both road and tunnel lighting promulgated by the CIE, and to a lesser extent in the many national recommendations covering these fields.

9.1 Roads

9.1.1 Lighting Criteria

A road lighting installation should provide an adequate degree of visual reliability for all road users, visual reliability being defined as the ability to

Table 9.1 Lighting criteria determining visual reliability

Components of Visual Reliability	Lighting Criterion		
	Level	Uniformity	Glare
Visual performance	Average road surface luminance L_{av}	Overall uniformity U_o	Threshold increment TI
Visual comfort	Average road surface luminance L_{av}	Longitudinal uniformity U_l	Glare control mark G

continuously select and process that part of the visual information presented necessary for the safe control of a motor vehicle. This can only be obtained if both visual performance and visual comfort are satisfactory (see Chapter 1), both of which are known in turn to be influenced by the three lighting criteria listed in table 9.1.

Although the influences of level, uniformity and glare will be discussed under separate headings, it should be appreciated that these criteria are in fact to some extent interrelated: for example, a higher lighting level permits of a lower uniformity for the same overall quality (see Chapters 2 and 3).

Lighting level

The lighting level should be specified in terms of the luminance needed on the road surface. With increasing luminance the contrast sensitivity of the eye increases (Sec. 1.2) and visual performance improves. The lighting criterion actually employed is the average luminance of the road surface between approximately 60 m and 160 m in front of the driver, this being the area representative of the background against which objects must be detected for most driving speeds.

The average road-surface luminance also influences the degree of visual comfort that a road lighting installation is able to afford the road user (Sec. 1.3).

The average road-surface luminance values to apply for the various categories of road (see next chapter) usually range from 0.5 cd/m^2, for roads with traffic of very low density and limited speed, to 2 cd/m^2 for roads where the density and speed of the traffic are high.

A distinction should be made between roads in dark and those in bright surroundings. In the latter case, where for example a road runs through a shopping area with brightly lighted display windows or close to bright advertisement signs, a higher average value of road-surface luminance is required than for roads in dark surroundings, e.g. most rural roads. This is because the light from the bright surroundings interferes with the normal adaptation state of the eye.

For moderately bright surroundings it is advisable to double the level normally recommended for the road lighting, but it should be borne in mind that much of the extra light needed will often in fact be provided in the form of stray light from the bright surroundings themselves.

For roads in both rural and urban areas the immediate surrounds of the road should also be adequately lighted. This is because obstacles on the road itself are frequently seen with the surround, be it footpath or whatever, as background. A well-lighted footpath will also make it easier for a motorist to spot a pedestrian about to cross the road.

A general lack of information concerning the reflection properties of footpaths makes it impossible to recommend a suitable luminance level for this. It should in general be adequate, however, if the average illuminance on the footpath is made not less than 50 per cent of that on the same width (say 5 metres) of adjacent carriageway.

It goes without saying that in purely pedestrian areas and residential areas where vehicular traffic is severely limited, the lighting needs of the pedestrians should predominate. (The lighting of residential and pedestrian areas is discussed in Chapter 16.)

Uniformity

Light from the road surface surrounding a dark spot on a road is scattered within the eye and interferes with the image of that spot focussed on the fovea. This can lead to an important loss of perceptibility for an object seen against the area of low luminance, and it is for this reason that the concept of overall luminance uniformity has been introduced. The overall uniformity (U_o), which is defined as the ratio of the minimum to the average road-surface luminance (Sec. 1.2), has a pronounced influence on visual performance, and for perception to remain acceptable it is recommended that U_o be no lower than 0.4.

From the point of view of visual comfort, the negative effect of a continuous sequence of bright and dark spots on the road in front of a driver can be limited by specifying a minimum value for the longitudinal uniformity (U_l). This is defined as the ratio of the minimum to the maximum road-surface luminance on a line parallel to the road axis running through the observer position (Sec. 1.3). Values of around 0.7 (Sec. 3.2) are generally recommended as a minimum. Such values will guarantee an adequate degree of visual comfort but do not, as is sometimes supposed, lead to a completely uniform impression of the road surface brightness pattern. Since visual comfort is relatively less important on minor roads, a minimum value for U_l of around 0.5 is acceptable for such roads.

The luminance gradient, viz. the rate of change of road-surface luminance with distance covered, also has an important influence on visual comfort (Sec. 3.2.2). A reduction in the luminaire spacing, if not compensated by a change in the value of U_l, could lead to an unacceptably high gradient. Research has shown that for a spacing of 25 m the U_l value must be some 10 per cent higher than for a spacing of 45 m. Lower values of U_l can be tolerated for much wider spacings (Sec. 3.2.2).

Since, as is illustrated in figure 9.1, there is no correlation between overall uniformity (U_o) and longitudinal uniformity (U_l), each must be assessed on its own merits.

a

b

Figure 9.1 Photographs taken in a road lighting model showing the absence of correlation between overall uniformity (U_o) and longitudinal uniformity (U_l): (a) Poor U_o and U_l (b) Poor U_o but good U_l.

Glare

Glare can take either of two forms. Glare that impairs the vision is known as disability glare, while glare that induces a feeling of discomfort is termed discomfort glare. Schreuder (1972) has shown that there is a certain correlation between these two glare forms, but that this correlation is so weak that each must be assessed individually.

Disability glare. The common measure for the degree of disability glare is the relative threshold increment, TI (Sec. 1.2.3). This can be calculated, in per cent, from the equivalent veiling luminance (L_v) and the average road-surface luminance (L_{av}) using the formula

$$TI = 65 \frac{L_v}{L_{av}^{0.8}}$$

for values of L_{av} lying between 0.05 cd/m^2 and 5 cd/m^2.

The threshold increment will vary continuously due to the changing position of the driver relative to the luminaires of the road lighting installation. So long as this variation is not too great, the variation itself will cause no disturbance, and it is sufficient to specify a top limit for TI.

The longitudinal position of the observer at which TI will be a maximum is dependent upon the screening angle of the vehicle's roof (figure 9.2). This angle has been standardised by the CIE for the purposes of glare evaluation at 20° above the horizontal. The value of TI will generally be greatest for that observer position where a luminaire appears just inside this angle (figure 9.3). It is advisable, therefore, to check the value of the threshold increment for this observer position.

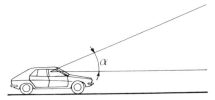

Figure 9.2 The screening angle α for a driver is the angle between the horizontal and the maximum upward line of sight.

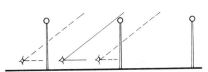

Figure 9.3 Position of observer, with a luminaire appearing just inside his field of view, at which TI (disability glare) is a maximum.

The value of TI has a considerable influence on the visual performance of a road user (Secs. 2.3 and 2.4). It is, of course, impossible for a road lighting installation to have a value of TI equal to zero, but values just below 10 per cent are quite feasible, and for main roads such values will limit the loss of visual performance due to glare to a tolerable extent. Of course, for minor roads higher TI values can be tolerated.

Discomfort glare. Discomfort glare is specified with the aid of the glare control mark (G), which is calculated from certain luminaire and installation characteristics (Sec. 1.3.3): the higher the value of G, the less will be the glare and the higher the visual comfort of the road user.

Satisfactory control of discomfort glare will be achieved with a G value greater than 6, but a G value of 7 or more is to be recommended. For minor roads G can, from a safety point of view, be somewhat lower, say around 5.

For roads with relatively bright surroundings the actual adaptation luminance of the eye is higher than would be expected from the average road-surface luminance. This has the effect of lessening the influence of both disability and discomfort glare: the higher the adaptation luminance, the less the glare influence. This means that for roads with surroundings that are bright relative to the road surface, TI can be higher and the glare control mark G lower than normal.

Visual guidance

The three lighting criteria so far mentioned (lighting level, luminance uniformity and glare) depend on the lamp luminaire combination used, the reflection properties of the road surface, and the geometry of the road and the luminaire arrangement. A fourth criterion is the extent to which the combination of these features of the installation serve to guide the road user through the scene ahead.

Good visual guidance makes it easier for the road user to see and correctly interpret the run of the road ahead of him and distinguish the boundaries of the traffic lane he is in and the points of intersection of this with other lanes or roads. Visual guidance, together with the other lighting criteria already mentioned, therefore goes a long way to ensuring an acceptable degree of visual reliability for the road user.

Of primary importance with regard to visual guidance are the design of the road, the road markings, and the guidance provided by the run of possible crash barriers. The guidance so provided should be strengthened by the lighting. Lighting level, luminance uniformity and glare control apart, a lighting installation can be made to provide good visual guidance – during the day as well as by night – by paying special attention to the alignment and arrangement of the luminaires such that these clearly indicate any changes important to a driver in the situation ahead. The systematic use of light sources of differing colour appearance can also contribute greatly to the quality of the visual guidance obtained (Sec. 15.2.1).

In many cases, the visual guidance provided by the lighting installation will in fact dominate. Even so, it is not possible, as it is with the other lighting criteria, to derive a figure to describe the degree of success in this direction.

9.1.2 Official Recommendations

There are basically three ways currently in use of presenting road lighting recommendations. First there is the so-called luminance approach. This method is in fact a direct practical 'translation' of the basic considerations covered in Part One of this book. The CIE 'Recommendations for the Lighting of Roads for Motorized Traffic' (CIE, 1977a) is an example of where such a method is employed.

Then there is the method originating from the British Standards Institution, 'Lighting for Traffic Routes' (BSI, 1974), which specifies the amount and distribution of the light from the luminaires together with the geometry of the installation for various road widths. This method is referred to here as the recepies method.

Finally, there is the method that uses the illuminance on the road surface as the basis for specifying road lighting quality. An example of a recommendation that adopts this approach is that prepared by the Illuminating Engineering Society of North America (AINSI/IES, 1977).

Luminance method

The requirements that road lighting installations must satisfy in order to provide visual reliability and so promote a smoothly moving and safe traffic pattern are dependent upon the intensity, speed and composition of the anticipated traffic system, and the layout of the road concerned; these factors considered in combination determine the road category.

Table 9.2 lists the five categories of road covered by the CIE in its recommendations (CIE, 1977a) while table 9.3, taken from the same recommendations, summarises the main quality criteria that can be expressed numerically. It will be noted, as mentioned in the previous section, that the brightness of the road surrounds relative to the road surface itself also plays a role with regard to the values recommended for L_{av}, G and TI. The values given are maintained values, that is to say an appropriate maintenance factor has to be applied to allow for the inevitable fall in light output between maintenance operations (see Chapter 11).

The recommended values given in table 9.3 are those applicable to dry-weather conditions. The CIE states that it is not possible, as yet, to give additional quantitative requirements relating to wet conditions. This is because of the general lack of information concerning, amongst other things, the reflection properties of wet road surfaces (Sec. 8.1.2). It is nevertheless desirable, of course, that the luminance uniformity on a wet surface be as good as that on a dry one, and ways in which this can as far as possible be achieved will be described in Chapter 10. Until more information on wet road conditions is available, and assuming that quality losses during wet weather

Table 9.2 Road classifications according to CIE

Category of road	Type and density of traffic*	Types of road	Examples
A		Road with separated carriageways, completely free of crossings at grade complete access control	Motorways Express roads
B	Heavy and high speed motorized traffic	Important traffic roads for motorized traffic only, possibly separate carriageways for slow traffic and/or pedestrians	Trunk roads Major roads
C	Heavy and moderate speed motorized traffic** or Heavy mixed traffic of moderate speed	Important, all purpose, rural or urban roads	Ring roads Radial roads
D	Fairly heavy mixed traffic of which a major part may be slow traffic or pedestrians	Roads in city or shopping centers, approach roads to official buildings and areas, where motorized traffic meets heavy slow traffic or pedestrians	Trunk roads Commercial streets Shopping streets etc.
E	Mixed traffic of limited speed and moderate traffic density	Collector roads between residential areas (residential streets) and A- to D-type roads	Collector roads Local streets etc.

mixed traffic | motorized traffic

* In cases where the road lay-out is below standard for the considered type and density of traffic, it is advised to install a lighting quality higher than the recommended one. In cases, however, where the road lay-out is considered to be above standard for the expected traffic density, a slight decrease of the lighting quality may be economically justified.

** Speed limit approx. 70 km/h.

150

Table 9.3 *Recommended (maintained) values of lighting parameters for road lighting (according to CIE Publication No. 12/2, second edition, 1977)*

Category	Surrounds	Luminance level	Uniformity ratios		Glare restriction	
		Average road surface luminance L_{av} (cd/m²) ≥	Overall uniformity ratio U_o ≥	Lengthwise uniformity ratio U_l ≥	Glare control mark G ≥	Threshold increment TI (%) ≤
A	any	2			6	10
B 1	bright	2		0.7	5	10
B 2	dark	1			6	10
C 1	bright	2	0.4		5	20
C 2	dark	1			6	10
D	bright	2		0.5	4	20
E 1	bright	1			4	20
E 2	dark	0.5			5	20

have not been restricted by the use of special rough (porous) road surfaces, an overall uniformity for wet roads lighted by conventional installations of 0.15 to 0.2 would seem a reasonable target to aim at (CIE, 1977a and 1980).

As a safeguard against inadequate lighting of the surrounds of a road, the same CIE recommendation states that it is desirable that a stretch some 5 m in width beyond the carriageway be illuminated to an illuminance not less than 50 per cent of that on the adjacent 5 m of carriageway.

Various national illuminating engineering societies have already started to adapt their recommendations to bring them more in line with the 1977 revised edition of CIE Publication 12. (See survey of recommendations at the end of this section.)

Recepies method

The Code of Practice for Road Lighting (Part 2) – the recepies method of the British Standards Institution (BSI, 1973 and 1974) – deals with the lighting of all-purpose traffic routes and presents a simple-to-use system for designing road lighting installations. (The Australian Standard (SAA, 1973) presents a system modelled along similar lines.)

The British Code uses a two-class classification of luminaires (cut-off and semi-cut-off, BSI 4533 – see Sec. 7.2.1) in which the classification is done

Design spacings (m)

Light distribution class (BS 4533)	Height (m)	Minimum lower hemisphere flux	Effective road width W (m)										
			11	12	13	14	15	16	17	18	20	22	24
cut off	10	12000 lm	33	33	33	31	29	27	26	24	22		
	12	20000 lm			40	40	40	40	37	35	32	29	26
semi cut off	10	12000 lm					44	44	42	40	36		
	12	20000 lm							53	53	52	47	43

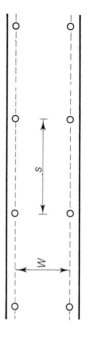

Figure 9.4 Table and sketch from the Code of Practice of the British Standards Institution giving design spacings (s) for cut-off and semi-cut-off luminaires employed in the opposite arrangement at various effective road widths (W).

152

according to the luminaire light distribution. The Code gives a table of design spacings for both classes of light distribution, for each of six basic lighting arrangements, two mounting heights (10 m and 12 m) and for a range of effective road widths. The minimum luminous flux of the luminaire in the lower hemisphere is also specified for each mounting height.

An extract from the table, together with one of the six accompanying sketches defining the lighting arrangement adopted is given in figure 9.4.

It should be noted on passing that the glare arising from the use of the two BS classes of luminaire at the two mounting heights specified in the table is more than that allowed by the CIE recommendations for category A, B and C roads.

It is stated in the British Code that the design spacings given in the table 'are such as to give an installation of high quality and good uniformity of brightness on a road of moderate skid resistance'. Nevertheless, the quality variations obtained with different luminaires belonging to a particular class can be considerable, even for one and the same road surface. It has in fact been found (Van Bommel, 1978b), in a test in which the overall quality of the lighting was judged on the basis of revealing power (a measure of visibility probability described in Sec. 2.3), that some BS semi-cut-off types of luminaire can be used at spacings wider than those recommended in the Code for other semi-cut-off luminaires without lowering the overall quality of the lighting. In some cases, the recommended spacing can be increased by 20 per cent or more, and even greater increases are possible if luminaires not belonging to one of the two BS light distribution classes are employed.

From what has been said above, it would seem that the recepies method, as described here, although very easy and convenient to work with for those having no access to more sophisticated design tools (such as, for example, the computer) is not really flexible enough to ensure a truly energy and cost-effective design for all luminaire types.

Design flexibility can, however, be achieved using a recently introduced design approach (described in Sec. 13.3 under the heading 'Performance Sheets') that has been developed to satisfy the requirements layed down in the CIE recommendations mentioned above under the heading Luminance Method.

Illuminance method
Prior to 1965, the year when the first International Recommendations for Road Lighting based on luminances was published by the CIE (1965), most national codes were based on illuminance. At the time of writing, some national codes still use the average illuminance and illuminance uniformity on the road surface as the sole basis for specifying road lighting quality. An example of such an illuminance oriented code is that published by the Illuminating

Engineering Society of North America (ANSI/IES, 1977). The Roadway Lighting Committee of this Society is, however, now considering going over more to a luminance oriented code (Ketvertis, 1979).

Since, as was demonstrated in Part One of this book, it is luminance rather than illuminance that is important in determining the visual performance and comfort of a road user, it follows that any road lighting design based on illuminance instead of on luminance will result in unpredictable variations in these criteria.

One could question whether or not the simplified road classification system used for luminance designs is precise enough to avoid giving the same degree of unpredictability in quality criteria that is present with the illuminance concept. But it has in fact been shown (Van Bommel, 1978a) that the quality variations are much greater with designs based on illuminance than for those employing the luminance concept and based on the classification system of dry road surfaces.

Table 9.4 Relative standard deviations of variations of the quality parameters U_o and U_l of lighting installations designed on the basis of equal illuminance uniformity and equal luminance uniformity (based on classification system) respectively

Standard deviation of:	U_o (%)	U_l(%)
Illuminance designs	20.1	41.7
Luminance designs	9.7	11.3

As regards the relative magnitudes of these variations for specific lighting criteria, these are given in table 9.4. The table is based on a study of 1620 installations and lists the standard deviations found in the variation in the uniformity criteria U_o and U_l. The variation in U_o is seen to be approximately twice as great with illuminance designs as with those based on luminance, and the variation in U_l some four times greater with illuminance than with luminance.

9.1.3 Survey of Official Recommendations

Table 9.5 provides a survey of the various official road lighting recommendations currently in force (the recommendations for areas classified as purely residential are not included). In nearly all cases, both the lowest and the highest values of the three quality parameters level, uniformity and glare corresponding to the least important and most important categories of road are given.

154

Country and literature reference	Level and range specified	Uniformity and range specified	Glare and range specified	Comments
CIE 1977	L_{av} 0.5–2 cd/m²	U_o 0.4 U_l 0.5–0.7	TI 10–20% G 4–6	
Australia (SAA 1973)	L_{av} 0.7 cd/m²	U_o 0.33 U_l 0.25		On the basis of a recepies oriented method (see Sec. 9.1.2)
Austria (LTAG 1979)	L_{av} 1 –2 cd/m²	U_o 0.4 U_l 0.5–0.7	luminaire class c.o. – n.c.o.	
Belgium (NBN 1979)	L_{av} 2 –4 cd/m²	U_o 0.4 U_l 0.6–0.7 U_t 0.30–0.35	luminaire class c.o. – n.c.o.	
Denmark (DEN 1979)	L_{av} 0.5–2 cd/m²	U_o 0.25–0.4 U_l 0.6	D 0.075–0.11 G 6–7	
Finland (NBR 1979)	L_{av} 0.5–2 cd/m²	U_o 0.4 U_l 0.6	TI 8–15% G 4–5	
France (AFE 1978)	L_{av} 0.5–2 cd/m²	U_o 0.4 U_l 0.5–0.7	G 4–6	
West Germany (DIN 1979)	L_{av} 0.5–2 cd/m²	U_o 0.3–0.4 U_l 0.4–0.7	luminaire class c.o. – n.c.o.	
Hungary (MSZ 1976)	L_{av} 0.25–2 cd/m² E_h 3–24 lux	U_{Ea} 0.2–0.3		
Italy (AIDI 1974)	L_{av} 0.5–2 cd/m²	U_o 0.4 U_l 0.5–0.7	TI 5–20% G 5–7	

155

Country (standard)	Illuminance / Luminance	U_o / U_{Eo}	U_l / U_{Ea} / U_t	Glare / luminaire class	Remarks
Japan (JSR 1967)(JIS 1969)	E_h 7–15 lux; L_{av} 0.5–2 cd/m²	U_o 0.5		luminaire class c.o. – n.c.o.	U_o value on the basis of standard geometrics
Netherlands (NSvV 1974)	L_{av} 1–2 cd/m²	U_o 0.2–0.4	U_l 0.5–0.7	G 5–6	
North America (AINSI/IES 1977)	E_h 4–22 lux	U_{Eo} 0.33			to restrict glare a minimum mounting height for the different luminaire classes is specified
Norway (VSV 1975)	L_{av} 0.5–2 cd/m²	U_o 0.4		luminaire class c.o. – n.c.o.	
Poland (PKNM 1976)	L_{av} 0.5–2 cd/m²	U_o 0.4		luminaire class c.o. – n.c.o.	
Sweden (NRA 1975)	L_{av} 0.5–2 cd/m²	U_o 0.4	U_l 0.6	D 0.12	higher glare value permitted for less important road categories
Spain (MOP 1964)	E_h 9–24 lux; L_{av} $\geqslant$ 0.8 cd/m²	U_{Eo} 0.35–0.65; U_o 0.50	U_{Ea} 0.20–0.35; U_t 0.33	L_l 2.10^4 cd/m²	
Switzerland (SEV 1977)	L_{av} 0.5–4 cd/m²	U_o 0.2–0.4	U_l 0.5–0.7	TI 10–20 %; G 5–6	
U.K. (BSI 1974)					recepies oriented method (see Sec. 9.1.2)
U.S.S.R. (BCH 1975)	L_{av} 0.4–1.6 cd/m²		U_a 0.20–0.33	TI 15 %	

The key to the table is as follows:

L_{av} average road-surface luminance

E_h average horizontal illuminance on the road

U_o overall luminance uniformity (L_{min}/L_{av})

U_l longitudinal luminance uniformity ($L_{min}/L_{max})_l$

U_t transverse luminance uniformity ($L_{min}/L_{max})_t$

U_a absolute luminance uniformity (L_{min}/L_{max})

U_{Eo} overall illuminance uniformity (E_{min}/E_{av})

U_{Ea} absolute illuminance uniformity (E_{min}/E_{max})

TI threshold increment

G glare control mark

D (no name) $L_v/(L_{av} + L_v)$ where L_v is the veiling luminance

L_l luminaire luminance in the direction of the observer for angles between $60°$ and $90°$ from the downward vertical

c.o.   cut-off $\quad\Big\}$ according to CIE, 1965 – see table 7.3.

n.c.o.   non-cut-off

9.2 Tunnels

9.2.1 Lighting Criteria

From the lighting point of view, a road tunnel can be divided into the four zones shown in figure 9.5.

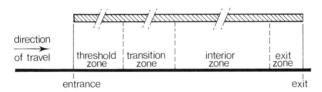

Figure 9.5 Tunnel zones distinguishable for lighting purposes.

The lower parts of the tunnel walls serve, as does the road surface itself, as a background against which objects in the tunnel are seen. The walls also constitute a large and important part of the field of vision, and their brightness influences the adaptation state of the eye. For these reasons, the lighting of the walls, at least as far as their luminance is concerned, is just as important as that on the road surface.

Threshold zone

In order that possible obstacles in the first part of a tunnel – the so-called threshold zone – be clearly visible to a driver approaching the tunnel entrance, the lighting level in this zone must be relatively high. The actual level required is dependent upon the magnitude and distribution of the luminances outside the tunnel that go to determine the eye adaptation state of the approaching driver.

The threshold zone luminance needed can be determined from the so-called equivalent adaptation luminance at the stop decision point – this being the point situated a safe driver stopping distance in front of the tunnel (Chapter 5). Figure 9.6 gives the safe driver stopping distance as a function of vehicle speed and road gradient.

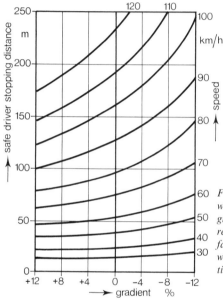

Figure 9.6 *Safe driver stopping distance in the wet, as a function of vehicle speed and road gradient. The values, which include a 1 s driver reaction time, are based on a clear road surface and tyres exhibiting an average amount of wear. (Taken from the German recommendations for tunnel lighting.)*

For practical design purposes the equivalent adaptation luminance at the stop decision point can be approximated by 1.5 times the average luminance in a conical field of vision subtending an angle of 20° at the eye of the observer at this point and centred on a horizontal viewing direction in the direction of travel. For tunnels in the planning stage, this average luminance can be approximated by comparison with existing tunnels where the lighting conditions and surroundings are similar (Van Bommel, 1980).

At the time of writing, more accurate methods are being considered for calculating the equivalent adaptation luminance from the luminance distribution in the field of view of a driver at this point (see Chapter 5).

Figure 5.8 of Chapter 5 gives the required threshold zone luminance as a function of the outside adaptation luminance, which for practical purposes can be considered as being equal to the equivalent adaptation luminance at the stop decision point.

The total length of the threshold zone must be equal to the safe driver stopping distance plus some 15 metres, this extra length being needed to create a background against which a possible obstacle at the end of the stopping distance will be silhouetted.

Transition zone

After the threshold zone comes the transition zone. Here the lighting level can be gradually reduced until at the end of the zone it is equal to the minimum value required in the tunnel interior. The permissible luminance reduction rate in the transition zone can be defined by Schreuder's curve given earlier in figure 5.10. (The way in which this curve is applied in practice is fully explained in Sec. 5.2.2.) By way of example, figure 9.7 shows the rate of decrease in luminance permitted in the transition zone together with the required threshold zone and interior zone luminances for a tunnel where the speed limit is 80 km/h. In practice, the luminance decrease can be achieved in a series of steps, the maximum permitted decrease between steps being in the ratio 3 : 1.

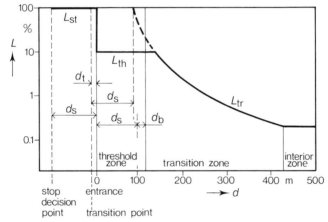

Figure 9.7 Permitted rate of luminance decrease in the transition zone of a tunnel where a speed limit of 80 km/h is in force. The luminance L_{th} in the threshold zone equals 10 per cent of the equivalent adaptation luminance L_{st}. Also shown, are the safe driver stopping distance ($d_s = 100$ m), the distance between the transition point and tunnel entrance ($d_t = 10$ m), and the distance ($d_b = 15$ m) needed to create a suitable object background.

Interior zone

The fact that it is always more dangerous to travel through a tunnel than along an open road means that higher lighting levels are required for the former than for the latter.

159

The longer the tunnel, the more time there is available for adaptation and the lower the lighting level in the interior zone may be. Recommended luminance levels for this zone range from $5\,\mathrm{cd/m^2}$ to $20\,\mathrm{cd/m^2}$. Where, because of the low level involved, it is not possible to use a continuous line of luminaires, the luminaire spacing should be made such as to avoid producing disturbing flicker (figure 5.14, Sec. 5.5).

Exit zone
As visual adaptation from a low to a high level of luminance takes place rapidly, and because a gradually increasing level of adaptation is experienced as the tunnel exit is neared due to daylight penetration, there is no need for special exit zone lighting.

9.2.2 Official Recommendations

Threshold zone
Most codes and recommendations on tunnel lighting, including the International Recommendations for Tunnel Lighting of the CIE (CIE, 1973), are based on the results of the research carried out by Schreuder, which are dealt with in detail in Chapter 5 of this book. With regard to threshold zone luminance, Schreuder's curve giving this as a function of the outside adaptation luminance (figure 5.8) is generally translated into the requirement that the former should be at least 10 per cent of the latter (BSI, 1971 – Great Britain; DIN, 1972 – Germany; IES, 1972 – USA; CIE, 1973; AFE, 1978 – France). The CIE recommendations add that, for economic reasons in particular, the 10 per cent value can be reduced to 7 per cent provided certain requirements are complied with, e.g. that there is a speed restriction in force or that a minimum distance between following vehicles is observed.

Since the various codes and recommendations make differing assumptions regarding the level of the outside adaptation luminance, it follows that the threshold zone luminance resulting from the application of the above 10 per cent rule can vary.

The present CIE recommendations are based on an outside adaptation luminance of $8000\,\mathrm{cd/m^2}$ for tunnels in flat, open country. For tunnels where special constructions are used to lower this luminance, and for tunnels in mountainous regions and built-up areas, the CIE gives the following formula from which the adaptation luminance (L_a) can be calculated

$$L_a = \frac{L_{d1} + 2L_{d2} + 3L_{d3}}{6}$$

where L_{d1}, L_{d2} and L_{d3} are the luminances in a conical field of view with an included angle of $20°$ centred on the direction of travel along the approach road to the tunnel at the following distances from the tunnel entrance in metres

$$d1 = 5\ V$$
$$d2 = 3\ V$$
$$d3 = V$$

where V is the speed limit in km/h. For example, at the speed often quoted earlier of 80 km/h, $d1 = 400$ m, $d2 = 240$ m and $d3 = 80$ m.

As was mentioned earlier, the CIE is considering recommending in future that the luminance at the beginning of the threshold zone be determined on the basis of the outside adaptation luminance at the so-called stop decision point (Sec. 9.2.1 and Chapter 5).

The outside adaptation luminances used in the British Code (BSI, 1971) as a basis for the 10 per-cent rule, range from 8000 cd/m² for 'unobstructed sunshine' via 2000 cd/m² for 'bright day' to 500 cd/m² for 'average overcast'. The threshold zone levels quoted in the German DIN recommendations are based upon an outside adaptation luminance that varies according to the speed limit in force, and hence the position of the stop decision point, from 3000 cd/m² to 6500 cd/m² (for 50 km/h and 100 km/h, respectively) for tunnels in the open country and from 2000 cd/m² to 4500 cd/m² (for 50 km/h and 100 km/h, respectively) for tunnels in built-up areas or in the mountains. The North American recommendations on tunnel lighting (IES, 1972) give no values for the outside adaptation luminance but state that at positions near to the tunnel entrance the adaptation luminance is determined by the average luminance in the visual field.

The French recommendations (AFE, 1978) propose that the outside adaptation luminance be determined using the CIE formula given above. The recommendations suggest possible values: 8000 cd/m² for underwater tunnels where the sky is bright, and for mountain tunnels whose entrances are surrounded by snow; 4000 cd/m² for tunnels in urban or mountainous areas, for tunnels where 25 per cent of the surrounds is of high luminance, and for tunnels whose entrances face south-east or south-west; 1000 cd/m² to 2000 cd/m² for tunnels where no high luminances enter the field of view on approach or where the entrance is surrounded by rocks or trees or by tall buildings such that it never receives direct sunlight.

Two national recommendations on tunnel lighting that are not based on the work of Schreuder, are those of Japan and Switzerland.

The Japanese recommendations (HRFJ, 1979) concerning threshold zone luminance are based mainly on research carried out in Japan by Narisada and Yoshikawa. (The results of this research were published in English – Narisada and Yoshikawa, 1974 and Narisada, 1975 – see Chapter 5.) Although the values quoted in the Japanese recommendations are lower than those recommended by the CIE, Narisada, concludes that had the same assumptions been made as in Schreuder's investigation concerning certain visibility requirements, such as contrast of critical object and presentation time of this, then the differences would have been minimal. For example, accepting a critical object contrast of 20 per cent, a presentation time of 0.1 s and a probability of correct perception of 75 per cent (the values accepted by the CIE and which form the basis of the investigations described in Chapter 5), gives a corrected Japanese value for the threshold luminance at an outside adaptation luminance of 4000 cd/m² and a speed limit of 100 km/h of 350 cd/m²; the ratio of threshold zone luminance to outside adaptation luminance is then close to the 1:10 recommended by the CIE.

Finally, the Swiss recommendations (SEV 4024, 1968) are based on research carried out by Mäder and Fuchs (1966).

The critical object employed by these workers was larger than that employed by Schreuder (and on which the CIE recommendations are based) – viz. 17′ × 34′ as opposed to 7′ × 7′. Consequently, the value recommended for the threshold zone luminance (L_{th}) is somewhat lower. The actual value is given by the formula

$$L_{th} = 0.15 \, L_a{}^{0.9}$$

where L_a is the outside adaptation luminance. For example, for an L_a of 4000 cd/m², L_{th} becomes 262 cd/m².

In most recommendations, those of the CIE included, the minimum length recommended for the threshold zone is based on the safe driver stopping distance. The CIE, in a recent technical report (CIE, 1979b), states that the threshold zone should be made equal to this distance plus a length needed to provide an object background. (Values of safe driver stopping distance can be read off from figure 9.6.)

Although the CIE, in its present recommendations (CIE, 1973), states that the luminance should be reasonably constant over the full length of the threshold zone, it is now considering whether or not the level should be allowed to progressively decrease slightly, the argument being that a driver's adaptation luminance also decreases as the tunnel entrance is approached (see Chapter 5). It is for this reason that the Japanese recommendations allow such a decrease in the last stretch of the threshold zone.

Transition zone

Many lighting codes and recommendations also draw on the research done by Schreuder (and described in Chapter 5) for the lighting of the transition zone. However, the manner in which Schreuder's luminance reduction curve is applied varies from recommendation to recommendation. This can be illustrated by considering how the CIE and German recommendations are carried out in practice.

It is stated in the CIE recommendations that the zero point of the curve's time or distance axis should be made to coincide with the end of the threshold zone, and the 100 per cent value of the luminance at this point put equal to the outside adaptation luminance (see figure 9.8). The German recommendations also place the zero point of the curve at the end of the threshold zone, but put the 10 per cent value of the luminance at this point equal to the luminance prevailing in the threshold zone (see again figure 9.8). The authors' proposal for the position of Schreuder's curve (shown in figure 9.8 by the dotted curve) is explained in Chapter 5.

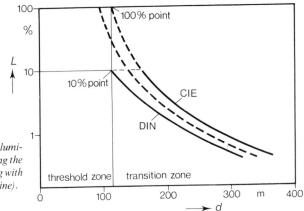

Figure 9.8 Application of the luminance reduction curve showing the CIE and DIN proposals along with that of the authors (dotted line).

The CIE further states that the luminance reduction curve may be approximated by a stepwise broken curve, provided that none of the steps in luminance is greater than 3:1.

The Japanese and Swiss recommendations regarding lighting in the transition zone are based on their own investigations carried out by Kabayama (1963) and by Mäder and Fuchs (1966) respectively. It has been shown, however, that these investigations and those of Schreuder (on which the CIE recommendations are based) lead to similar results (figure 5.10).

In some recommendations the requirement for the lighting level in the transition zone is given in terms of the permissible luminance reduction from one stretch of defined length to the next, the length being expressed in terms of travel time. The British Code (BSI, 1971), for example, says that the changes in the levels of luminance should not be greater than a reduction to one half of the previous level (i.e. luminance reduction ratio 2 : 1), and that about 3 s of

163

travel time should be allowed at each luminance level before a further reduction occurs. (In practice this recommendation allows much the same rate of reduction as that proposed by the authors.) The North American recommendations (IES, 1972) allow a very rapid transition from the luminance in the threshold zone to that in the interior zone, the reduction in luminance in fact being accomplished in one 10:1 step for which the travel time should be 4 s or more.

Interior zone

The CIE recommends that the tunnel interior be lighted to give the following daytime luminances
10 cd/m² to 20 cd/m² in urban tunnels
  5 cd/m² to 10 cd/m² in rural tunnels
  3 cd/m² to 5 cd/m² in very long tunnels or tunnels with an appropriate speed restriction in force or where the traffic density is light
The above values are more or less in line with those appearing in the majority of national recommendations.
In the interior zone, with its relatively low lighting level, it may be decided to employ discontinuous rows of luminaires. This can lead to problems with disturbing flicker unless certain flicker frequencies are avoided. The CIE states that the critical frequency range to be avoided at all costs is 2.5 pps to 15 pps (light pulses per second). (The range of critical frequencies decreases as the length of exposure to the flicker is reduced, i.e. as the tunnel length becomes shorter – Sec. 5.5.)

Exit zone

The CIE, in common with most national recommendations, does not give a requirement for the lighting of the tunnel exit zone, the adaptation of the eye from low to high luminances being very rapid. A warning is, however, given against allowing direct sunlight to enter the exit.

Night-time lighting

The lighting level in the interior zone of a tunnel can be reduced at night to between 2 cd/m² and 5 cd/m², but the luminance of the open road at the tunnel entrance or exit should not be less than one-third of the actual tunnel luminance prevailing (CIE, 1973). (For the transition from normal road lighting to no lighting, see Sec. 15.2.2.)

Switching

In order not to present a hazard to a driver, the change in lighting level from the daytime maximum to the night-time minimum should be performed in steps of no greater than 3 : 1 (CIE, 1973).

Chapter 10

Bad-weather Lighting

A road lighting installation designed to fulfil the necessary quality requirements for dry weather conditions can nevertheless allow bad, and thus dangerous, visual situations to arise due to the presence of rain or fog. The question of exactly what quality should be aimed at during adverse weather conditions is still being investigated, as is the question of how to design an all-weather installation possessing the necessary quality requirements. It is, however, possible to briefly summarise the most important factors having a bearing on road lighting quality under adverse conditions and to suggest some practical guidelines for arriving at good quality bad-weather lighting.

10.1 Wet Weather

In wet weather, a certain amount of the light directed towards the road from a road lighting installation never reaches it: some is lost through absorption in the rain drops and some is scattered by these in all directions. The resultant loss of visibility, although troublesome while the rain lasts, is however usually mild compared with that caused by the wetting of the road surface which may remain wet long after the rain has stopped.

The influence that a wet road surface has on the visibility obtained with a given lighting installation is a direct result of the changed nature of that surface's reflection properties. The average luminance of a surface increases when it becomes wet and the overall and longitudinal uniformities decrease. The main problem is this decrease in uniformity, especially the decrease in overall uniformity, U_o.

As was mentioned when dealing with road surfaces (Sec. 8.1.2), a generally agreed system for describing the reflection properties of wet surfaces does not yet exist as it does for dry surfaces. This means that it is still difficult to design an installation such that it will have a given wet-weather specification. Certain practical guidelines can, however, be given with regard to the best choice of road surface, luminaire light distribution and lighting arrangement.

10.1.1 Road Surfaces

As a general rule, it is logical to expect that quality losses in the lighting occurring during wet weather can be kept low by the use of road surfaces that show only small changes in their reflection properties when they become wet: in practice, the lower the specularity of the wet surface, the better the overall and longitudinal uniformities will be. This is illustrated in figure 10.1 (Van Bommel, 1976b and 1976c), where for a twin-central arrangement on a dual carriageway with central reservation these two uniformities have been plotted as a function of the corrected specular factor $S1'$ (See Chapter 8) of the wet road surface. The increase in uniformity with decrease in $S1'$ is clearly shown. Road surfaces having a low degree of specularity under dry conditions generally tend to have a low specularity under wet conditions also. This is shown

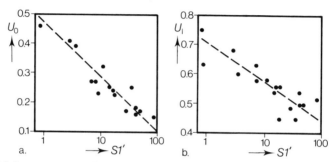

Figure 10.1 Decrease in overall and longitudinal uniformity with increase in wet surface specularity. Each point represents the average uniformity value obtained using twenty-five different luminaires employed in the twin-central arrangement on a dual carriageway with central reserve.

in figure 10.2, where the corrected specular factor $S1'$ (wet) has been plotted as a function of $S1$ (dry) for some twenty different road surfaces.

This relationship between dry and wet surface specularities is again evident from a study done by the Scandinavian Traffic Lighting Research (STLR, 1978); although this study also indicates that the correlation can be masked when the surfaces compared exhibit different degrees of wear.

This may explain why Frederiksen and Sørensen (1976) found no clear correlation between the specularity of a road surface in the dry and wet conditions. It also means that other factors apart from the specularity when dry have an important bearing on the specularity of a surface when wet.

One such factor is the texture depth of the surface. Sørensen (1975) showed that the tendency is for the specularity of a wet road surface to decrease with increase in texture depth, (figure 10.3). This is in line with what was said earlier (Sec. 8.1) concerning the reflection characteristics of a wet surface being dependent mainly upon the surface's macro-structure, of which the texture depth is a measure.

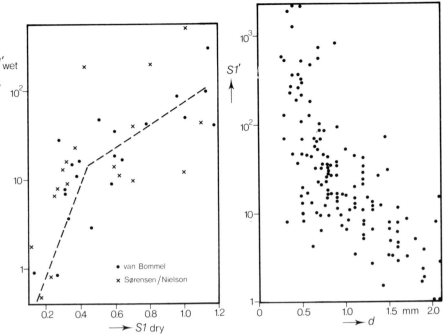

Figure 10.2 Road surface for which dry surface specularity (S1) is low will generally exhibit a low wet surface specularity (S1') also.

Figure 10.3 Decrease in wet surface specularity with increase in texture depth for chip-coated asphalt and asphalt concrete road surfaces. (Sørensen)

The composition of a road surface, the prevailing climate, and the traffic load all serve to directly influence the texture of that surface and thus its dry and wet-weather specularities. So far as the composition is concerned, some guidance for arriving at that needed to secure a low wet-weather specularity can be obtained from the draft Danish Code for Road Lighting (DEN, 1976). The Code indicates (for conditions prevailing in Denmark) the probable W-class (W for wet – see also Sec. 8.1.2) corresponding to certain road surface compositions. The relevant part of the Code is summarised in table 10.1.

It should be emphasized that although, as will be shown in the following two sections, the light distribution of the luminaires and the type of lighting arrangement also have an influence on the wet-weather road lighting quality, the most important influence is without doubt that exerted by the nature of the road surface itself.

10.1.2 Light Distributions

Light coming from the luminaires of a road lighting installation and reflected specularly by the wet road toward a road user will produce bright patches

Table 10.1 Relationship between composition of the road surface and W-class summarized from the draft Danish Code for Road Lighting

Asphalt concrete	Max. grain size less than 10 mm	W 4
	Max. grain size from 10–18 mm	W 3/W 4
Surface dressing	With coated chippings, max. grain size more than 10 mm	W 2/W 3
Pervious asphalt		W 1/W 2
Concrete	–	> W 4
	With surface texture improved by brushing, cutting etc.	< W 4
General	Surface treatments (in good state)	< W 2

on its surface, so reducing the luminance uniformity. Luminance uniformity can be improved by employing luminaires that have a reduced luminous intensity in the directions in which specular reflection towards an observer can occur, whilst throwing as much light as possible in other directions.

Van Bommel (1976b and 1976c), following this line of reasoning, derived a luminaire figure of merit – the so-called cross factor (CF) – with which it is possible to describe the suitability for use under wet conditions of any given luminaire. He showed that such a figure of merit could best be defined as the intensity ratio I_c/I_l where (figure 10.4):

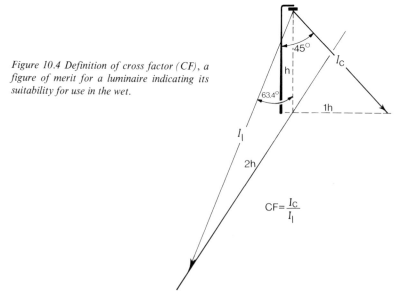

Figure 10.4 Definition of cross factor (CF), a figure of merit for a luminaire indicating its suitability for use in the wet.

I_c = the luminous intensity of the luminaire in the direction $45°$ ($\tan^{-1} 1$) from the downward vertical in the plane at right angles to the road axis (crosswise direction); and

I_l = the luminous intensity of the luminaire in the direction $63.5°$ ($\tan^{-1} 2$) from the downward vertical in the plane parallel to the road axis (longitudinal direction).

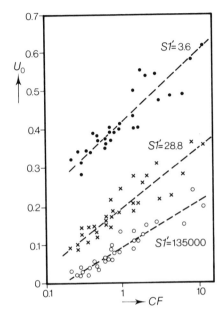

Figure 10.5 Calculated increase in overall uniformity U_o with increase in cross factor CF for three types of wet surface. Valid for a twin-central lighting arrangement on a dual carriageway with central reserve.

Figure 10.5 gives the results of calculations in which the overall uniformity (U_o) has been calculated for different wet road surfaces and for luminaires having different CF values. The figure clearly illustrates that the higher the CF of a luminaire, the better will be the uniformity achieved with it on different types of wet surface, and thus the better also will be the wet-weather quality of the lighting installation as a whole.

10.1.3 Lighting Arrangements

Another way of improving the wet-weather quality of a road lighting installation is to ensure that what bright patches do occur are evenly distributed over the entire road surface. A noticeably greater improvement in luminance uniformity can be brought about in this way by employing a staggered or an opposite arrangement of the lighting columns rather than a single-sided or a twin-central pole arrangement.

169

How great the improvement will be is, of course, largely dependent upon the road layout, the type of road surface and the type of luminaire employed. Van den Bijllaardt (1976) compared the single-sided with the staggered arrangement, and the opposite with the twin-central arrangement for a single road cross-section, for four wet road surfaces and for two different luminaires. He showed that on average there was a 50 per cent improvement in the overall uniformity for the staggered arrangement over the single-sided one, and a 35 per cent improvement for the opposite over the twin-central.

With all arrangements, the wet-weather road lighting quality can be improved by bringing the luminaires as far as possible towards the centre of the carriageway (i.e. by increasing the luminaire overhang), so increasing the amount of light shed across the road towards the area where no specular reflection towards the road user can occur.

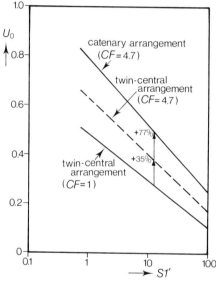

Figure 10.6 Calculated improvement in overall uniformity with decrease in wet surface specularity S1' for three cross factor/lighting arrangement combinations on a two-lane dual carriageway. The results are based on an examination of twenty different wet road surfaces.

A very good wet-weather road lighting installation is the catenary installation. The typical catenary arrangement, with its closely spaced luminaires (See Sec. 15.1.2), is better able to spread the light evenly over the road surface. Catenary luminaires are, moreover, characterised by their large cross factors. A catenary luminaire can have a CF value of 10 or more, compared with the CF of less than 1 of most conventional luminaires.

Van Bommel (1976b) determined the improvement possible in overall uniformity for a dual carriageway with central reserve by using catenary luminaires (with CF = 4.7) instead of the same number of conventional lu-

170

minaires (with CF = 1) placed in the twin-central arrangement (see figure 10.6). For a wet road surface having a corrected specular factor $S1'$ equal to 10, for example, the total improvement is 77 per cent. By also considering the purely theoretical case of catenary luminaires employed in the twin-central arrangement, he showed that an improvement of some 35 per cent is attributable simply to the higher cross factor possessed by the catenary luminaire.

10.2 Foggy Weather

The poor visibility which may occur on a lighted road during foggy weather can be attributed to light being absorbed and scattered by the fog-producing water droplets suspended in the atmosphere. Some of this light is lost, and some is reflected towards the road user in the form of a bright veil of fog that obscures details on the road ahead. (In vehicle lighting, this is commonly known as the 'white-wall effect'.) Whilst nothing can be done regarding the loss of light, there are certain measures that can be taken to see that the light scatter towards the road user is reduced to a minimum.

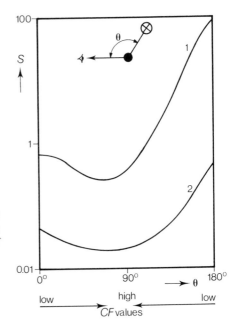

Figure 10.7 Relative degree of light scatter S in fog (curve 1 dense fog, curve 2 light fog) as a function of the angle θ between direction of light incidence and direction of scatter. (Foitzik/Zschaeck)

The degree to which fog scatters light in a given direction is dependent upon the angle that this direction makes with the incident light beam. The relative degree of scatter has been measured by Foitzik and Zschaeck (1953) for both dense and light fog conditions, figure 10.7. From the figure it can be seen that scatter drops to a minimum as the angle θ approaches 90°. It therefore follows, if the direction of scatter is taken as being towards the eyes of a road user (viz. along a line parallel to the road axis), that this condition for minimum scatter will be readily fulfilled by any luminaire having a high cross factor, the major portion of whose light output will be across the road and so at right angles to the line of sight.

As mentioned above, the highest cross factor is that possessed by the catenary luminaire. The catenary installation is therefore ideally suited for use in situations where fog is frequently encountered, especially since the visual guidance obtained with such an installation is also excellent.

Chapter 11

Lighting Maintenance

A road lighting installation will continue to operate efficiently only so long as it is well maintained. Furthermore, poor maintenance can allow the harmful effects of lamp lumen depreciation and lamp failure, and luminaire fouling and depreciation to bring about an intolerable deterioration in the quality of the lighting.

Some deterioration in quality is of course inevitable, even with a well maintained installation, and this deterioration will be greatest just prior to maintenance being carried out. The task facing the lighting designer is that of fixing the maintenance period and the initial lighting level given by the installation such that the lighting level and quality can never fall below specification. At the same time, the installation must not be made too expensive to maintain or too heavy on energy.

The factors influencing the rate of deterioration, the maintenance operations that must be carried out to restore the lighting quality, and the advantages and disadvantages of different maintenance programmes, particularly with regard to costs, are briefly dealt with in this chapter.

11.1 Deterioration

11.1.1 Lamp Failure

Tolerances in manufacture alone make it impossible for a lamp to be given a precisely defined life or to predict when a particular lamp will fail. Moreover, with the discharge lamps commonly employed in road lighting, lamp life is influenced by several factors not controlled by the lamp manufacturer, e.g. type of ballast, ambient temperature, operating voltage, voltage fluctuations, frequency of the on/off switching, burning position, severity of mechanical vibrations, and so forth. Variations in lamp life also occur according to the wattages of the lamps considered.

No generally valid information regarding lamp failure times can therefore be given. What the lamp manufacturer can supply, however, are so-called mortality curves for specified lamp types operated under laboratory controlled conditions. Each curve indicates the percentage of lamps of a particular type still functioning as the number of burning hours is increased. The number of burning hours elapsed before 50 per cent of the lamps have failed is called the average lamp life.

As will be pointed out later in this chapter, the concept of average lamp life defined in this way is of little practical use in the field of road lighting. More important, particularly with regard to cost, is a knowledge of the number of burning hours corresponding to a much lower mortality rate. It is for this reason, and also because the reliability of a mortality curve decreases rapidly with increase in the number of burning hours, that mortality curves are often given for up to a maximum of only 10000 burning hours.

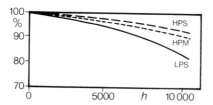

Figure 11.1 Mortality curves for high-pressure sodium (HPS), high-pressure mercury (HPM) and low-pressure sodium (LPS) lamps, where h is the number of burning hours.

Some typical mortality curves, *obtained under laboratory controlled conditions*, are given in figure 11.1. These curves should not be used for reference purposes. They do, however, show the trend for the discharge lamps employed in road lighting; namely, very few failures occurring during the first few thousand burning hours.

11.1.2 Lamp Lumen Depreciation

Lamp lumen depreciation, that is to say the fall in light output during lamp ageing, occurs with most lamps. How great this is at a given moment in the life of a lamp is dependent not only upon the type and wattage of the source: many other factors, such as ambient temperature, voltage fluctuations, burning position, severity of mechanical vibrations, and so forth, also play an important role. Therefore, as is the case with lamp failure, no generally valid information on depreciation under actual operating conditions can be given.

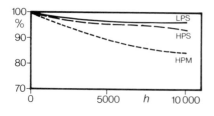

Figure 11.2 Lamp lumen depreciation curves, where h is the number of burning hours.

174

Lamp manufacturers can, however, supply lumen depreciation curves for the various lamps (type and wattage) in their programme, as obtained under laboratory controlled conditions. Some typical lamp lumen depreciation curves are given in figure 11.2. They show the trend that depreciation with high and low-pressure sodium lamps, especially the latter, is lower than with high-pressure mercury lamps.

11.1.3 Luminaire Fouling

By far the most serious loss in light output in road lighting is that brought about by the accumulation of dirt on the light emitting surfaces of the luminaires. This luminaire fouling may also unfavourably alter the light distribution of the luminaires.

The severity of the fouling will, of course, depend on the type of luminaire involved and the amount of pollution locally present in the atmosphere. Closed luminaires afford much better protection against dirt accumulation, and also corrosion (see below), than do open luminaires, provided of course that the sealing is adequate (see also Sec. 7.1.6). Nevertheless, even with closed luminaires fouling can, in heavily polluted areas, easily reduce the lumen output by as much as 20 to 30 per cent per year, while in pollution-free areas a reduction of 5 to 10 per cent per year is quite possible.

11.1.4 Luminaire Depreciation

As a luminaire ages there will occur a gradual loss of reflectivity of reflectors and mirrors perhaps accompanied by a loss of transparency of the (refractor) bowl due to the onset of corrosion and discolouration. These effects are, of course, also dependent upon the type of luminaire involved and the materials used in its construction. The light output lost due to luminaire depreciation cannot usually be regained by cleaning, and for closed luminaires a 1 per cent drop in light output per year should be reckoned with as being the average for a luminaire of good quality.

11.2 Maintenance Operations

The total effect of the various factors influencing depreciation mentioned above has to be estimated in advance when a new road lighting installation is being designed; only then will there be any guarantee that the lighting will remain within specification throughout the life of the installation.

Of course, the overall extent of the effect of depreciation will largely depend upon the nature of the maintenance operations and the frequency with which these are carried out. For example, the longer the interval between cleaning periods, the greater will be the effect of dirt accumulation on lumen output and thus the higher the lighting level required for the installation when new. Lamps can be replaced individually, as they burn out, or all the lamps in the installation can be replaced at one time, after a reasonable number of burning hours, whether they have failed or not. The latter system is commonly referred to as group replacement. It is usual, during the group replacement operation, to clean the luminaires and inspect the electrical and mechanical connections. Sometimes a combination of group replacement and individual spot replacement (for failed lamps at critical locations) is adopted.

Group replacement (combined with luminaire cleaning and inspection) generally offers the most advantageous solution as far as cost and energy saving is concerned. The fall in light output of the luminaires due to lamp lumen depreciation and luminaire fouling can in this way be kept relatively low; as a consequence, the required 'initial' lighting level given by the installation will be comparatively close to the maintained level.

Detailed information on the method of determining the best group replacement period is given in CIE Technical Report No. 33 (CIE, 1977b). (The number of burning hours corresponding to this period is sometimes referred to as the economic life of the lamps.) An interval between successive group replacement (and cleaning) operations of from 6000 to 8000 burning hours (viz. 1.5 to 2 years) is common. In heavily polluted areas it sometimes pays to add an extra cleaning operation between group replacements rather than shorten further the replacement interval.

Even where systematic maintenance is employed, the lighting level can easily drop by as much as 20 to 30 per cent by the time lamp replacement is due. A maintenance factor corresponding to the drop anticipated (e.g. 0.8 for a 20 per cent drop and 0.7 for 30 per cent) should therefore be applied when setting the initial lighting level to be given by the installation.

Chapter 12

Cost and Energy Considerations

There are indications (Scholz, 1978), comparing the cost savings obtained from the reduction in the number of road accidents with the operating costs of road lighting, that good efficient road lighting is economically feasible (see also Chapter 1). It is nevertheless, of course, desirable that costs and energy consumption be kept to the minimum consistent with the maintenance of adequate road lighting quality.

Many factors play a role in determining the costs and energy effectiveness of a road lighting installation. There are, for example, factors such as lamp and luminaire type, type of control gear employed, the pole type and arrangement, luminaire mounting height, the reflection properties of the road surface, the location of the electricity supply cables (i.e. above or below ground), and the type of switching.

The relative importance given to each of the above factors will be dependent upon the local circumstances prevailing regarding such things as the cost of labour, and the cost and availability of materials and energy. There are therefore no generally applicable rules for minimising on costs and energy, especially not for complicated non-standard installations such as catenary and high mast. (Van den Bijllaardt, et al., 1975, did a cost evaluation of catenary installations valid for the circumstances prevailing in the Netherlands.) However, where conventional installations are concerned it is possible to indicate certain general tendencies with regard to the way in which cost and energy consumption is influenced by the three factors: lamp type, luminaire type, and road surface type.

12.1 The Influence of Lamp Type

12.1.1 Luminous Efficacy and Price

The luminous efficacy of a lamp has a dominating influence on the energy consumption and consequent running costs of a road lighting installation employing that lamp. Figure 12.1 shows the increase in lamp efficacy that continuous research has brought about over the years.

Another important difference between the various lamp types is that of price.

Low-pressure and high-pressure sodium lamps are more expensive than high-pressure mercury lamps, and these in turn are more expensive than tubular fluorescent lamps.

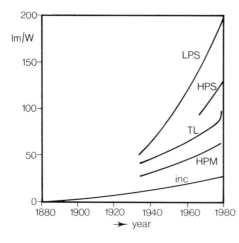

Figure 12.1 Rough schematic showing the increase in luminous efficacy of various lamp types over the last hundred years. (TL = tubular fluorescent, inc = incandescent)

12.1.2 Comparing Costs and Energy Consumption

The total costs and energy consumption of road lighting installations employing different lamp types can easily be compared using a system described by Van Bommel (1978b).

First, the maximum pole spacing needed to satisfy the uniformity requirements for the installations considered is determined for a given pole height and luminaire light distribution. (Use can be made here of the performance sheets described in Chapter 13.) The luminance level obtained with this spacing for a specific lamp is unlikely to exactly equal the level specified. But this level can be theoretically matched by employing the appropriate hy-

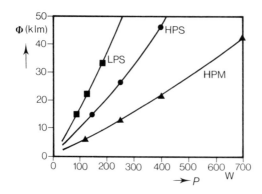

Figure 12.2 Hypothetical relation between luminous flux and lamp wattage based on existing lamp types.

pothetical lamp type, viz. one giving the required lumen output. The relation between lamp wattage and luminous flux for the hypothetical lamps (figure 12.2) is based on currently available figures for existing lamps.

The geometric data (spacing, height) so obtained form, together with the cost and energy data, the input to a cost comparison calculation where the prices of the hypothetical units (lamp/luminaire/control gear combination) are obtained by interpolation between the prices given for existing types.

This system guarantees an honest comparison, since the lighting quality of the installations is near enough the same for the various lamp types. (The fact that hypothetical lamp types are considered means that the results obtained can also be used to point the way in the field of lamp development.)

Results using this system have been published for low-pressure sodium, high-pressure sodium and high-pressure mercury lamps (Fischer and Kebschull, 1978). (The tubular fluorescent lamp is only suitable for use in providing low-level lighting, and for this reason was not considered.) The input data were as follows:

Lighting arrangements: Single-sided, opposite, and twin central (figure 12.3).

Quality requirements: These were based on the CIE recommendations for category A and B_2 roads respectively (see table 9.3) and are specified in figure 12.3.

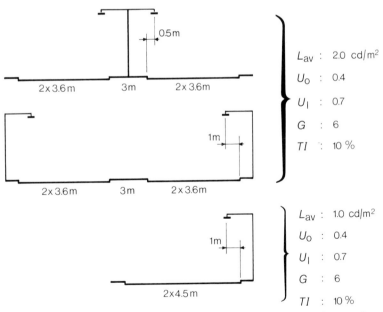

L_{av} :	2.0 cd/m²
U_0 :	0.4
U_1 :	0.7
G :	6
TI :	10 %

L_{av} :	1.0 cd/m²
U_0 :	0.4
U_1 :	0.7
G :	6
TI :	10 %

Figure 12.3 The lighting arrangements, road cross-sections and quality specifications for which cost evaluations have been made.

Luminaire light distributions: As given by CIE tight-glare-control luminaires belonging to one and the same quality class with respect to longitudinal and lateral distribution (viz. similar throw and spread).

Road surface reflection properties: Standard R surface of class 3 (with a Q_0 of 0.08).

Costs: Based on estimated average 'world' prices.

Group relamping – done after 8000 h for a labour cost per lamp of from 10 to 15 Dutch guilders, according to pole height.

Individual 'spot' relamping – done at the rate of 2 per cent per year for a labour cost per lamp of from 75 to 115 Dutch guilders, according to pole height.

Energy – the kWh price has been taken as 0.15 Dutch guilders.

(For more information on the cost data used, see original publication.)

Maintenance factor: 0.8.

The total annual cost per kilometre of installation (made up of the amortisation of the initial investment over 20 years, and energy and maintenance costs) is compared in figures 12.4 and 12.5 as a function of luminaire mounting height.

The symbols employed in these two figures indicate the positions of existing lamp types.

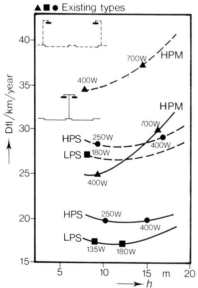

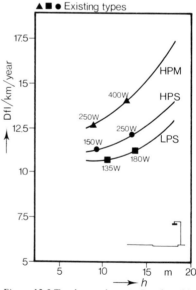

Figure 12.4 Total annual cost in Dutch guilders (Dfl) per kilometre of lighting the dual carriageways specified in figure 12.3, as a function of luminaire mounting height.

Figure 12.5 Total annual cost in Dutch guilders (Dfl) per kilometre of lighting the two-lane road specified in figure 12.3, as a function of luminaire mounting height.

It can be seen that the low-pressure sodium lamp always results in the lowest total cost whichever of the three lighting arrangements is employed; although this does not mean, of course, that the initial investment will also be lowest for this lamp. On dual carriageways more favourable results, especially for the low-pressure sodium lamps, are achieved with the twin-central rather than the opposite arrangement, which is only to be expected. Where, because of the need to provide at least some degree of colour rendering, the low-pressure sodium lamp cannot be used, the high-pressure sodium lamp gives the best solution from the point of view of minimising costs; the use here of the high-pressure mercury lamp is clearly hardly justified on these grounds.

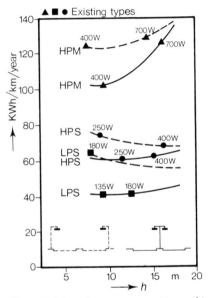

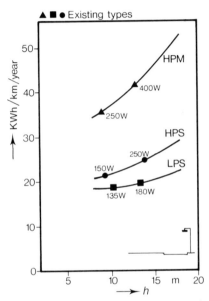

Figure 12.6 Annual energy consumption per kilometre corresponding to figure 12.4.

Figure 12.7 Annual energy consumption per kilometre corresponding to figure 12.5.

The energy consumption comparison shown in figures 12.6 and 12.7 clearly demonstrates the enormous energy saving possibilities offered by low-pressure sodium and, to a slightly lesser extent, high-pressure sodium compared with high-pressure mercury.

It is interesting to note that the lowest total costs are obtained at the lower mounting heights. For the two-lane dual carriageway the most economical pole height would appear to lie between 8 and 12 metres which is equal to slightly more than the width of one carriageway, while for the single carriageway the most economical pole height is 8 metres, which is again approximately equal to the carriageway width (below 8 metres the requirement with regard to U_o – viz. $U_o \geqslant 0.4$ – is no longer satisfied).

The tendency shown with regard to costs is also true with respect to energy conservation, with the exception of the low-pressure sodium and high-pressure sodium lamps used in the opposite arrangement.

These results disprove the often-held belief that savings on costs and energy consumption can always be obtained by employing greater mounting heights and consequently larger spacings.

Although the cost and energy figures derived from such a comparison will in practice vary from country to country, the trends indicated are in general true for most countries. In fact, throughout the world, conversion programmes from high-pressure mercury to low and high-pressure sodium are being pressed forward in an endeavour to improve the cost and energy effectiveness of road lighting installations (Spinelli, 1976 and 1978; and Claxton, 1978).

12.2 The Influence of Luminaire Type

Lamp types apart, the type of luminaire employed, specifically the light distribution of this, also has a considerable influence on both cost and energy consumption.

An important consideration in this respect is the luminaire's downward light output ratio (d.l.o.r. is the ratio of the downward luminous flux of the luminaire to the bare lamp flux). However, the use of a luminaire having a high d.l.o.r. but a poor light distribution necessitates employing a closer pole spacing in order to meet luminance uniformity requirements. A combination of a high d.l.o.r. and a good light distribution is in fact the only way of keeping costs and energy consumption to a minimum.

In order to satisfy the glare requirements laid down by the CIE for the higher categories of road (categories A, B and C_2 – see tables 9.2 and 9.3) the use of tight-control luminaires is called for. There are now such luminaires available, incorporating computer-designed optics, that in most cases permit of a spacing to mounting height ratio of about 4 to 1 whilst still satisfying the uniformity requirements for these categories of road; this in place of the 3 to 1 maximum generally achievable with earlier types. The d.l.o.r. of these improved luminaires is at least equal to that of the earlier types.

The increase in spacing brings with it a 25 per cent saving in the number of lamps, luminaires and masts. However, because the lamps employed at this increased spacing must have a higher luminous output, and therefore a higher wattage, the energy saving is somewhat less than would be expected, viz. between 10 and 15 per cent.

It is often thought that most non-tight-control luminaires (viz. CIE 1965 semi-cut-off or non-cut-off luminaires, which result in intolerable glare for the higher categories of road) offer tremendous cost savings. But this is not always the case. Sarteel (1975) made a cost comparison of installations in Belgium and came to the conclusion that it is not necessarily more expensive to employ CIE 1965 cut-off luminaires rather than semi or non-cut-off luminaires to satisfy the minimum requirements concerning level and uniformity. Indeed, in some cases it is even more economical to employ cut-off luminaires. Where the use of semi or non-cut-off luminaires does provide a cost saving, this is on average in the order of 15 per cent, which is a relatively small return for the inevitable loss of overall lighting quality caused by the increase in glare associated with these luminaires.

12.3 The Influence of Road Surface Type

Road surfaces are today chosen mainly on the basis of their mechanical strength, skid resistance and cost. It has been shown in Chapter 8, however, that the reflection properties of a given surface exert a considerable influence on the luminance obtainable with a given amount of light, thereby influencing also the total cost of the lighting installation.

The lighter the road surface (viz. the higher the average luminance coefficient, Q_0) the lower the luminous flux of the lamps needed to obtain the specified lighting level, the luminous flux needed being inversely proportional to the increase in the Q_0 value. Artificially brightened asphalt surfaces and concrete surfaces can have a Q_0 that is as much as 20 to 50 per cent higher than that possessed by normal asphalt surfaces. It is to be hoped, therefore, that such surfaces will find greater application in the future. (Surfaces with these high Q_0 values usually have a low degree of specularity, which is an extra advantage in connection with wet-weather quality – Chapter 10.)

Figures published for the situation prevailing in Belgium (Prevot, 1978) show that the annual cost of a lighting installation (one that is comparable with others in terms of lighting quality) is on average 17 per cent less for artificially brightened surfaces and approximately 25 per cent less for concrete surfaces compared with conventional asphalt surfaces. (The cost of the road surfaces themselves is not included in these figures.)

Chapter 13

Calculations

During the actual lighting design phase of a lighting project, a lighting engineer has to perform lighting calculations in order to arrive at solutions that will satisfy the relevant lighting requirements.

Since the performance of lighting calculations involving luminance is relatively time consuming, extensive use is now being made of computers. Universally applicable computer programs are available for this purpose, as will be shown in the last part of this chapter.

But for routine design work the computer is often not needed; provided, that is, that suitable information concerning the performance of the luminaires under various more or less standard conditions of use is available in an acceptable form. An example of how this performance information can be presented and used in manual design work will be given in this chapter.

Where lighting calculations have to be carried out for non-standard situations (e.g. exceptional road width, mounting height, or road curvature) and a computer is not available, use can be made of the photometric data (in graph form) of the relevant luminaires. The method of using these graphs for the purposes of calculating the average illuminances and luminances as well as the uniformities and glare values will also be outlined in this chapter.

13.1 Conventions

13.1.1 Calculation Points

The values obtained for calculated averages and uniformities, whether they be illuminances or luminances, are only of course meaningful if both the number and the position of the calculation points employed are specified.

It is generally accepted that the area on the road for which the calculations are performed – the calculation grid – should be representative of the road between 60 and 160 metres in front of the road user, which is the area of greatest interest to him. Test calculations have, in fact, shown that this requirement will be fully satisfied if the calculation grid covers at least one span (this being the area between successive luminaires on the same side of the road) and if this is placed roughly in the middle of the 100 metre stretch of road specified above.

The CIE (1976a) recommends that where the luminaire spacing does not exceed 50 metres there should be ten evenly spaced transverse rows of calculation points over this length, while for luminaire spacings greater than 50 metres the number of transverse rows should be such that the distance between two successive rows does not exceed 5 metres (figure 13.1).

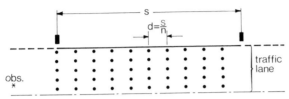

Figure 13.1 Calculation grid as proposed by the CIE: s = spacing; d = longitudinal distance between calculation points; n = number of transverse rows. For s ⩽ 50 m, n = 10; s > 50 m, n = smallest integer giving d ⩽ 5 m.

Finally, the CIE also recommends that there should be five points across each traffic lane, with one point positioned on the centre line of each lane. It is stated that where the uniformity is good – $U_o \geqslant 0.4$ – subsequent calculations may be based on three points instead of five.

13.1.2 Observer Position

The luminance and glare calculations made in road lighting must take into account the most probable position of the road user relative to the area of the road surface for which the calculations are made. This position, the so-called observer position, can be fixed by specifying the eye height of the observer above the road surface and his distances from the beginning of the calculation grid (longitudinal position) and the side of the road (lateral position).

Eye height
In the CIE publication referred to above it is recommended that eye height be taken as 1.5 metres, this being close to the average height of drivers' eyes above the road.

Longitudinal position
From what has been said above, it can be seen that the observer is assumed to occupy a position some 60 metres plus (depending on the width of the span involved) in front of the calculation grid.

There is, however, one exception to this rule, and this is in the case of the glare parameter *TI* (the threshold increment). This should be calculated for the longitudinal observer position giving the highest *TI* value (see Sec. 9.1.1). Since the screening angle formed by a car's roof is standardised at 20° (CIE, 1976b), *TI* will generally be at a maximum for that longitudinal position of the observer at which a luminaire appears just within this screening angle (see also figure 9.3); the longitudinal observer position at which this situation arises is therefore the one assumed to be most appropriate when calculating *TI*.

Lateral position

The CIE defines the lateral position of the observer according to the type of calculation being performed. For the calculation or measurement of all quality parameters other than longitudinal uniformity, the observer is considered as occupying a position one quarter of the carriageway width from the nearside kerb – the so-called Standard Lateral Observer Position. (For a single-sided, or asymmetrical, lighting arrangement there will in fact be two such observer positions, one for each direction of traffic flow.) Longitudinal uniformity, on the other hand, is calculated for the line passing through an observer positioned in the middle of each traffic lane and facing the direction of traffic flow.

In order to gain an insight into the influence that lateral observer position has on the values obtained for the various quality parameters, Van Bommel (1977) calculated these for different values of the former. The calculations were performed for single-sided, staggered and twin-central lighting arrangements and for a number of dry and wet road surfaces. The results obtained are summarised in figure 13.2, which shows the maximum deviation of the various parameters with change in the lateral position of the observer from the

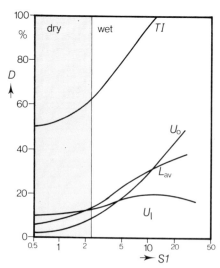

Figure 13.2 *The effect of departure from the Standard Lateral Observer Position on the four quality parameters indicated, as a function of road surface specularity S1, where D is the corresponding percentage deviation in the value of the parameter.*

186

Standard Lateral Position defined above, as a function of the specular factor $S1$. As can be seen, for dry road surfaces (shaded area in figure 13.2) the level and uniformity parameters (viz. L_{av}, U_o and U_l) show a deviation of no more than about 10 per cent for any departure from the standard observer position defined above. In other words, the error in calculating L_{av}, U_o and U_l for the Standard Lateral Observer Position alone rather than employing a separate position, as is done by the CIE for the calculation of U_l, would appear to be unwarranted, at least for dry-weather conditions. For the glare parameter (TI), however, the change in value with departure from the Standard Lateral Observer Position is quite considerable – as much as 60 per cent. Where glare can be critical, therefore, a check on TI at observer positions other than the Standard Lateral Position defined above, would seem advisable, even though not actually recommended by the CIE.

For wet road surfaces, it is apparent from this figure that observer position has a much greater influence on all four quality parameters. However, before any definite conclusions can be drawn from this concerning which is the best observer position to take for the purposes of establishing a convention – or whether indeed more than one position should be considered – further investigative work on wet-weather lighting would have to be carried out.

13.1.3 Average and Perspective Average

A road user sees the road ahead of him in perspective and gives weight to a luminance according to the apparent size of the area concerned: the closer the area is to him, the larger it will appear and the more influence it will have in comparison with other areas of equal real size farther away. In the average road surface luminance as normally calculated, the individual luminance values are not weighted according to the apparent size of the areas concerned, that is to say too little weight is given to the areas close to the observer position and too much to those farther away (figure 13.3a).

By distributing the individual calculation points uniformly over the per-

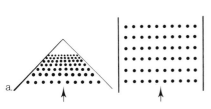 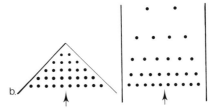

Figure 13.3 Average and perspective average. (a) Seen in perspective, the normal calculation grid gives undue weight to the concentration of points remote from the observer. (b) In this modified grid the points are so arranged that the correct weight is given to each when the grid is seen in perspective.

spective picture of the road surface (figure 13.3b) and calculating from these points, the perspective average road surface luminance is obtained.

As the road user travels along the road, his area of interest naturally moves along in front of him. Ideally, therefore, the average perspective luminance should be averaged for a large number of observer positions between one span and the next (figure 13.4). It has been shown, however, that the average of the perspective averages so obtained (termed the dynamic perspective average) is very close to the normal, unweighted (static) average calculated for a calculation grid that starts at the beginning of a span. It follows, therefore, that in practice the calculation of this unweighted static average will suffice.

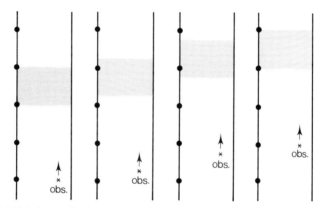

Figure 13.4 Calculating the dynamic average involves taking a number of observer positions between one span and the next.

13.2 Graphical Photometric Luminaire Data

Using the appropriate graphical presentations of the photometric luminaire data (Sec. 7.2.2), the point and average values of both illuminance and road surface luminance can be calculated in a straightforward way for any combination of lighting arrangement and road cross-section. The techniques employed are described below.

13.2.1 Illuminance

Illuminance at a point

The horizontal illuminance at a point on the road can be determined from the isolux diagram belonging to the particular luminaire type in use. A typical isolux diagram is shown in figure 13.5a. As can be seen, the grid of the diagram is specified in terms of the mounting height (h) of the luminaire, so making it valid for all mounting heights. The values of the illuminance contours are expressed as a percentage of the maximum illuminance given by the luminaire.

The relative illuminance produced by a single luminaire at a point can thus be read direct from the diagram; provided, of course, that the position of the point on the grid is defined in terms of h. The total relative illuminance at the point is then found by adding to this value the contributions made to the illuminance by the other luminaires. In practice it is usually sufficient to sum the contributions made by only three luminaires, viz. that nearest the point considered and one on either side of it in order to ensure a reasonably accurate result.

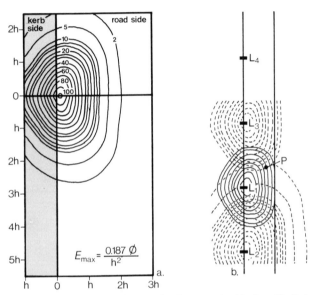

Figure 13.5 Illuminance at a point. This can be determined with the aid of the isolux diagram for the luminaire concerned (a) and a similarly scaled drawing of the installation in question (b) – see text.

In performing this summation the isolux diagram is used as follows. A copy of the isolux diagram made on a transparent sheet is layed over a plan of the road drawn to the same scale as that of the diagram itself, the origin of the diagram coinciding with the position of the first luminaire (figure 13.5b). The contribution made by this luminaire to the relative illuminance at the point is then read off. This procedure is repeated for the other two luminaire positions of interest, and the contribution made by all three luminaires summed.

It is important to ensure that the correct kerb-side/road-side orientation of the isolux diagram is maintained; for example, had the luminaires in the example illustrated been on the other side of the road, it would have been necessary to turn the isolux diagram through 180 degrees.

Having found the total relative illuminance at the point, all that remains is to convert this to an absolute illuminance value. This is done by multiplying the total relative illuminance by the maximum illuminance produced by one luminaire, the latter being defined on the diagram in terms of the luminous flux of the lamp used (Φ) and the mounting height employed (h) – e.g. $E_{max} = 0.187 \times \Phi/h^2$.

Average illuminance

The average illuminance for a defined area of calculation (for example, the area described in Sec. 13.1.1) may be found using

$$E_{av} = \frac{\Sigma E_p}{n}$$

where E_p is the illuminance at each point considered and n is the total number of points regularly distributed over this area.

But of course, this approach is very time consuming. For a straight road, however, the average illuminance can be quickly and easily calculated from the utilisation factor diagram for the particular luminaire in use.

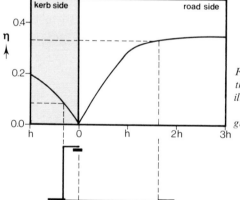

Figure 13.6 Utilisation factor diagram. For the road cross-section illustrated: η kerb side = 0.09, η road side = 0.33 giving η total = 0.42

The utilisation factor η of a road lighting luminaire is defined (See Chapter 7) as the ratio of the utilised flux, Φ_u (the flux reaching the road), to the luminous flux emitted by the lamp(s), Φ.

Thus

$$\eta = \frac{\Phi_u}{\Phi}$$

190

The value of η for a given road cross-section is read from the diagram as illustrated in figure 13.6. Knowing η, the average illuminance can be calculated using

$$E_{av} = \frac{\Phi_u n}{A} = \frac{n\,\eta\,\Phi}{w \cdot s \cdot n} = \eta\frac{\Phi}{w \cdot s}$$

where
η = utilisation factor of luminaire for the road cross-section considered
Φ = luminous flux of lamp installed in luminaire (lm)
w = width of road (m)
s = luminaire spacing (m)
n = number of luminaires
A = area of road (m^2)

13.2.2 Luminance

Luminance at a point
The luminance at a point on a road can be determined using the isoluminance diagram appropriate to the luminaire and road surface class considered. (Isoluminance diagrams are usually given for the four standard road surfaces corresponding to the four dry road-surface classes.) A typical isoluminance

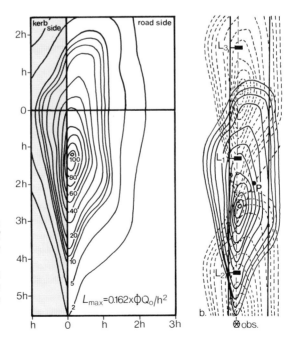

Figure 13.7 Luminance at a point. This can be determined with the aid of the isoluminance diagram for the luminaire concerned (a) together with a similarly scaled drawing of the installation in question (b) – see text.

$L_{max} = 0.162 \times \Phi Q_o / h^2$

191

diagram is shown in figure 13.7a. The grid of the diagram is specified in terms of the luminaire mounting height (h) to make the diagram valid for all heights. The isoluminance contours are specified as percentages of the maximum luminance.

Each diagram is calculated for an observer stationed in the vertical plane parallel to the road axis and passing through the luminaire (the $C = 0°$ plane) and a distance $10\ h$ from it. The way in which the diagram is used will thus depend on the actual position of the observer. There are basically two cases to consider: observer in-line with the row of luminaires, and observer outside the row.

1. Observer in-line with luminaires

Here the method of use is quite straightforward. First a scale plan of the road, drawn in terms of the mounting height of the luminaire, is made. A transparent copy of the appropriate diagram is then placed on this plan with the longitudinal axes of road and diagram parallel and with the centre point of the diagram on the projected position of the luminaire. The relative luminance at the point of interest is then read off.

By repeating this procedure for a second and third luminaire, and summing the results, it is possible to find the total relative luminance at the point (figure 13.7b).

Finally, the absolute value of the luminance at the point is found by multiplying the total relative luminance by the maximum luminance produced by one luminaire, the latter being defined on the diagram in terms of the luminous flux of the lamp used, the average luminance coefficient (Q_o) of the road surface, and the luminaire mounting height (h) – (e.g. $L_{max} = 0.089 \times \Phi \cdot Q_o/h^2$). Once again, it is important to maintain the correct kerb-side/road-side orientation of the diagram.

2. Observer positioned outside the row of luminaires

The luminance at a point on the road between an observer and the luminaire is dependent not only upon the light distribution of the luminaire but also upon the position of the point relative to the observer and the luminaire.

Conversely, the luminance at a point on the road behind the luminaire is almost entirely dependent upon the light distribution of the luminaire, and very little upon the position of the observer.

This means that the isoluminance diagram, for an observer in the $C = 0°$ plane, can be used here in the way described above with a fair degree of accuracy, provided the point of interest lies behind the luminaire. For a point between the luminaire and the observer, however, the diagram should be rotated such that its longitudinal axis is in line with the observer's position as marked on the plan of the road (figure 13.8); in this way a more accurate result

192

will be obtained. The relative luminance at the point is then read off from the diagram, and the luminance calculated as in case 1 above.

The method is accurate to within $\pm 10\%$ so long as the diagram is not rotated through an angle greater than $5°$; that is to say, the observer must be within a distance of $0.875\,h$ from the $C = 0°$ plane of the luminaire at the specified viewing distance of $10\,h$.

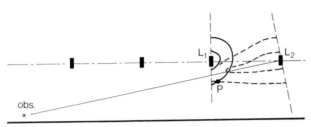

Figure 13.8 Determining the luminance at a point for an observer outside the row of luminaires – see text.

In cases where a more accurate calculation is needed or where the diagram would have to be rotated through an angle greater than $5°$, a new isoluminance diagram that takes into account the actual observer position should be constructed.

This can be done using a special isoluminance, or reference diagram (one that defines, for a particular road surface, the luminance distribution for a theoretical luminaire, the luminous intensity of which is constant for all angles of emission, figure 13.9) in conjunction with the isocandela diagram of the luminaire concerned. The most suitable isocandela diagram for this purpose is that in which the isocandela contours are projected on the road surface (see figure 7.10).

The new isoluminance diagram is made by multiplying values on the appropriate reference isoluminance diagram (viz. the one for the road surface class considered), positioned with its longitudinal axis pointing towards the observer position, by the values of the corresponding points on the isocandela diagram of the luminaire concerned. This will give the relative luminance at each of the various points on the new isoluminance diagram for the observer position considered.

Average luminance

Knowing the luminance values over a part of a road, the average luminance over the same area may be found from

$$L_{av} = \frac{\Sigma L_p}{n}$$

where L_p is the luminance at each point in a regular grid (for example, the

193

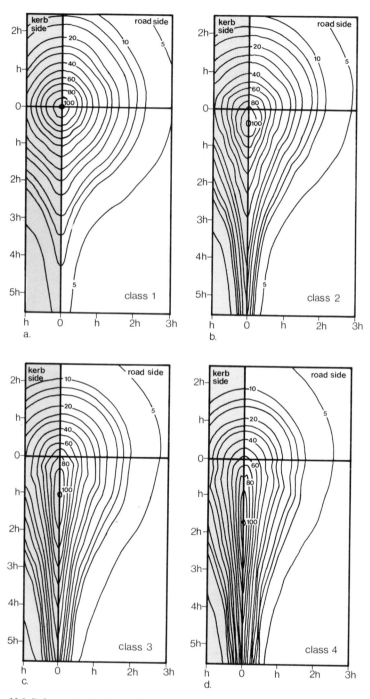

Figure 13.9 Reference isoluminance diagrams for four standard road surfaces for a theoretical luminaire, the luminous intensity of which is constant for all angles of emission.

194

calculation grid described in Sec. 13.1.1) and n is the total number of points considered. But of course, this approach is very time consuming.

The quickest way of calculating the average luminance of a straight road of infinite length is by using the luminance yield diagram (described in Chapter 7) appropriate to the luminaire and road surface class concerned. An example of a luminance yield diagram is given in figure 13.10. This gives the so-called luminance yield factor of the luminaire as a function of the transverse distance from the vertical plane through the luminaire parallel to the road axis for three positions of the observer; if the observer position for which the average luminance is wanted does not coincide with one of the positions defined in the diagram, the correct luminance yield factor may be found by interpolation (as illustrated in figure 13.10).

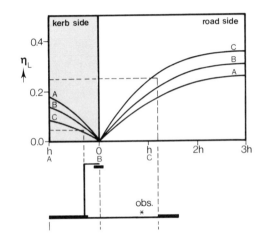

Figure 13.10 Luminance yield diagram for three observer positions relative to the row of luminaires: A – 1 h to the left, B – in line, C – 1h to the right. For the road cross-section and observer position illustrated:
η_L kerb side = 0.05
η_L road side = 0.25
giving η_L total = 0.30

Having determined the luminance yield factor for the road cross-section, observer position and road-surface class being considered, the average luminance L_{av} can be obtained from

$$L_{av} = \eta_L \frac{\Phi}{w \cdot s} Q_o$$

where η_L = luminance yield factor
   Φ = luminous flux of the bare lamp used in the luminaire (lm)
   w = width of road (m)
   s = luminaire spacing (m)
   Q_o = average luminance coefficient of road surface (cd/m²/lux)

This formula follows direct from the definition of luminance yield factor given in Sec. 7.2.2.

Luminance uniformity
Overall uniformity (U_o). The overall uniformity should be calculated as the ratio between the smallest of the calculated grid-point luminances and the average luminance.

Longitudinal uniformity (U_l). The longitudinal uniformity should be calcu-lated as the ratio between the smallest and the largest calculated luminances occurring in the calculation grid along the centre line of each traffic lane. The worst of the uniformities so calculated, for each direction of traffic flow, is then the longitudinal uniformity of the installation.

13.2.3 Glare

Threshold increment
With the aid of the luminous intensity distribution of a luminaire (as for example given in the isocandela diagram in equal-area zenithal projection, described in Sec. 7.2.2), and using the nomogram of figure 13.11 (Adrian, 1975) the threshold increment of an installation using such luminaires can be determined. This nomogram is a graphical representation of the formula for

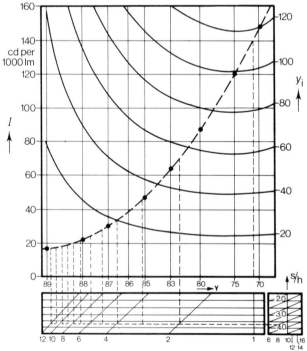

Figure 13.11 Nomogram used in determining the threshold increment (TI) for an installa-tion – see Appendix C (Adrian)

the equivalent veiling luminance given previously. The threshold increment (TI), in per cent, is obtained from the equivalent veiling luminance (L_v) and the average road surface luminance (L_{av}) using the formula given in Sec. 1.2.3:

$$TI = 65 \frac{L_v}{L_{av}^{0.8}}$$

Glare control mark
The logarithmic terms in the formula for the glare control mark given in Sec. 1.3.3 can be calculated fairly easily with the aid of a simple pocket calculator given the luminaire-dependent values together with the luminaire's photometric data. When the specific luminaire index (SLI) of a luminaire is given,

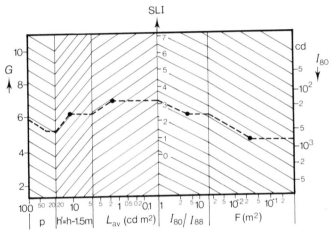

Figure 13.12 Nomogram used in determining the glare control mark (G) of an installation – see Appendix C.

the calculation can be simplified, for the formula for the glare control mark (G) then becomes (as described in Sec. 1.3.3)

$$G = SLI + 0.97 \log L_{av} + 4.41 \log h' - 1.46 \log p$$

where SLI = specific luminaire index
L_{av} = average road surface luminance (cd/m^2)
h' = reduced mounting height (m) (equals $h - 1,5$)
p = number of luminaires per kilometre

Alternatively, the glare control mark can be determined using the nomogram given in figure 13.12

13.3 Performance Sheets

For routine design work the relatively time consuming method of calculation based on the use of graphical photometric luminaire data can be avoided – even when no computer is available – if performance sheets such as those described in Chapter 7 are available for the luminaires considered.

As already mentioned in Sec. 7.2.2, performance sheets may take many different forms. However, each sheet should at least give the values of the photometric quality parameters (luminance level, luminance uniformity and glare ratings) for a specified road cross-section, lighting arrangement, luminaire-lamp combination and mounting height for various luminaire spacings and for the four standard road-surface reflection classes. (Figure 7.16 shows an example of a performance sheet that gives the values mentioned, plus illuminance values, for eleven different spacings and two mounting heights.)

A library of such performance sheets covering a range of typical road cross-sections and lighting arrangements (see figures 7.17 and 7.18) can be used to facilitate the design of road lighting installations. The first step, of course, is to narrow the number of sheets down to those depicting a road cross-section the same as the one for which the design has to be made. It is then simply a matter of going through these sheets one by one, and noting for each which of the precalculated lighting schemes (viz. pole arrangement, lamp-luminaire combination, mounting height and spacing), satisfies the quality requirements in force for the particular road-surface reflection class being considered. Should the road cross-section, the lighting arrangement, or both, differ from those contained in the library, it will then be necessary to interpolate between known values in order to arrive at the required design figures.

13.4 Computerised Calculations

Where a more detailed design analysis is required than can be provided by the precalculated performance sheets described above, the speed and accuracy obtainable with a computer are essential.

Illuminance, luminance and glare calculations made using photometric luminaire data presented in graphical form are rather tedious and time consuming. Such methods should therefore be seen more as a way of performing a limited number of spot checks on a tentative design proposal rather than as an aid in the design process itself.

The computer can be programmed to perform the basic illuminance and luminance calculations described in this chapter using just two equations

1. For illuminance at a point, E_p

$$E_p = \frac{I_{(C,\gamma)}}{h^2} \cos^3 \gamma$$

2. For luminance at a point, L_p

$$L_p = q_{(\beta,\gamma)} E_p = \frac{I_{(C,\gamma)}}{h^2} r_{(\beta,\gamma)}$$

where $I_{(C,\gamma)}$ = luminous intensity in the direction of the point as defined by the angles C and γ

h = luminaire mounting height

$q_{(\beta,\gamma)}$ = luminance coefficient of the road surface for the angle (γ) of light incidence concerned and angle (β) between vertical plane of observation and plane of light incidence

$r_{(\beta,\gamma)}$ = reduced luminance coefficient $(q \cos^3 \gamma)$ of the road surface for the angle (γ) of light incidence concerned and for the angle (β) between vertical plane of observation and vertical plane of light incidence.

The CIE recommends that tables giving the necessary intensities (termed I-tables) and reduced luminance coefficients (termed r-tables) be drawn up according to standard formats, the former to contain 1872 values in a defined C–γ grid (see figure 14.4) and the latter 396 values in a defined β–γ grid (see figure 14.6). Intensity and reflectance values for C–γ and β–γ combinations not appearing in the tables must, of course, be arrived at by a process of interpolation.

The glare equations needed by the computer programmer are those given in Sec. 1.2.3 and 1.3.3.

The computer programs employed for the calculation of the above illuminance, luminance and glare values can be structured to suit various input and output requirements and to give various degrees of sophistication. Two programs used for luminance calculations are described in CIE Technical Report No. 30. One of them is a standard program which is simple in use but which can only be used for straight road sections with relatively simple lighting geometries. The other program is a comprehensive one, and is able to deal with a greater number of situations concerning road and lighting geometry but is, for this reason, also more complicated to use.

Chapter 14

Measurements

The measurements carried out in connection with road lighting fall into two groups: laboratory measurements and field measurements.

Much of the information on which road lighting calculations are based is derived from the results of extensive photometric measurements conducted in specialised road lighting laboratories. The vast majority of these measurements are to do with determining the performance characteristics of luminaires and road surfaces; more specifically, the luminous intensity distribution of the former and the reflection properties of the latter. Although much of the equipment employed is nowadays fully automatic – from performing the actual measurements right through to printing out the results in a convenient form – the underlying measuring principles employed are basically simple. These principles are outlined in the first part of this chapter.

The second part of the chapter takes a look at field measurements. Field measurements made on a newly completed road lighting installation can be used to check that it is in line with recommended practice and that it gives the quality expected of it.

Further field measurements made at a later date will then reveal whether there is a need for maintenance, modification, or perhaps even replacement. Here, the quantities measured to check the quality characteristics of the installation are the road surface luminance and illuminance. Field measurements can also be employed as a means of assessing the reflection properties of road surfaces.

14.1 Laboratory Measurements

14.1.1 Luminous Intensity Distribution

A knowledge of the most important performance characteristics for a particular luminaire is derived from the luminaire's measured luminous intensity distribution. Strictly speaking, this can only be defined and measured for a point source, but for conventional road lighting luminaires the measurement inaccuracy will be negligible provided the optical path length of the measuring set-up is at least ten times the length of the light emitting surface of the luminaire. For narrow-beam floodlights still greater optical path lengths are required.

Of particular interest when measuring space is at a premium, is a technique that makes use of a unidirectional-sensitive photometer to shorten the path length needed. The way unidirectional sensitivity is achieved and utilised was first described by Frederiksen (1967). Languasco et al. (1979) described a recent modification to this technique.

Measuring techniques

Four fundamentally different measuring techniques are available for determining the complete luminous intensity distribution of a luminaire (figure 14.1):

1. The luminaire is rotated, generally about its longitudinal and transverse axes, while the photocell remains fixed in position (figure 14.1a).

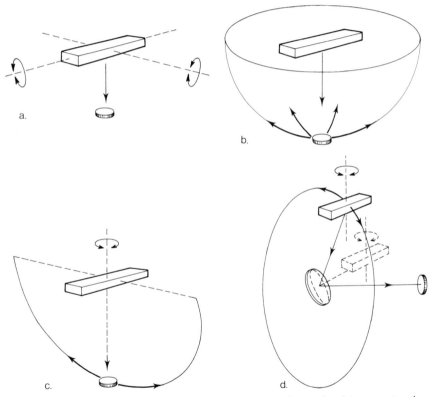

Figure 14.1 Schematic representation of the four basic principles employed in measuring the luminous intensity distribution of a luminaire: (a) Luminaire is rotated about two axes while photocell remains stationary (b) Luminaire remains stationary while photocell is moved over a hemispherical path (c) Luminaire is rotated about vertical axis while photocell travels along a semicircular path (d) Luminaire is moved relative to a moving mirror, which directs the light towards a stationary photocell.

2. The luminaire is fixed in position while the photocell is moved over a hemispherical path (figure 14.1b).
3. The luminaire is rotated about its vertical axis while the photocell is moved over a semi-circular path (figure 14.1c).
4. The luminaire is free to move, but in such a way that its normal operating position is at all times maintained, its light being directed toward the fixed photocell by a single rotating mirror or by a number of these (figure 14.1d).

With some lamps, particularly those of the discharge variety, both the luminous flux and the luminous intensity distribution of the lamp are dependent upon its burning position. The first of these measuring techniques, in which the burning position of the lamp changes as the luminaire is tilted, is therefore not generally suitable where such lamps are involved.

The second technique has the advantage over the first that the burning position of the lamp is fixed. This can, however, also be seen as a disadvantage, for it means that a correspondingly larger measuring space is needed in order for the photocell to move freely in three dimensions instead of being fixed. The stationary-luminaire technique is therefore really only suitable for use in large laboratories or in combination with a device that folds the optical measuring path.

The third technique referred to above can be seen as a logical improvement of the second. The photocell is moved, not over a hemispherical path, but over a semicircle in a plane passing through the luminaire's vertical axis, and the luminaire is rotated about its vertical axis whilst maintaining the normal burning position. In this way, the measuring path is long in one vertical plane only.

The fourth technique, which uses a mirror or mirrors rotating so as to direct the light from the moving luminaire towards a fixed photocell, has the important advantage of requiring a long measuring path in one direction only, viz. from mirror to photocell. A typical example of a measuring system employing this technique, which is the one generally adopted today, is illustrated in figure 14.2.

Presentation of results

The results of a luminous intensity distribution measurement can be presented in a concise and convenient way by making use of an appropriate system of coordinates. The three systems developed for use in road lighting, each of which is associated with a particular measuring technique, are the so-called $A-\alpha$, $B-\beta$ and $C-\gamma$ systems shown in figure 14.3. For conventional road-lighting luminaires, the $C-\gamma$ system is the one most favoured while for floodlights the $B-\beta$ system (and less often the $A-\alpha$ system) is the one generally employed.

Figure 14.2 A goniophotometer employing the principle illustrated in figure 14.1d. Through the inspection window can be seen the large rotatable mirror. The equipment in the foreground automatically controls the movement of the luminaire/mirror combination and produces a print-out of the luminaire's intensity distribution in the form of a polar diagram.

For general-purpose use, the measured luminous intensity values for the luminaire in question are often presented in the form of polar curves for each of a number of vertical planes (the C-planes of figure 14.3c) or in an isocandela diagram (for examples of both, see Chapter 7).

Where the measured luminous intensity distribution is to be used for computer-calculation purposes, the luminous intensity values must be given in a digital form in a luminous intensity table. The CIE (1973a) recommends that for conventional road-lighting luminaires the table should list intensity values for 1872 combinations of C and γ, the table to be layed out according to figure 14.4. Figure 14.5 shows part of such a table as drawn up for a specific luminaire.

In order that a single tabulated set of luminous intensity values may be used

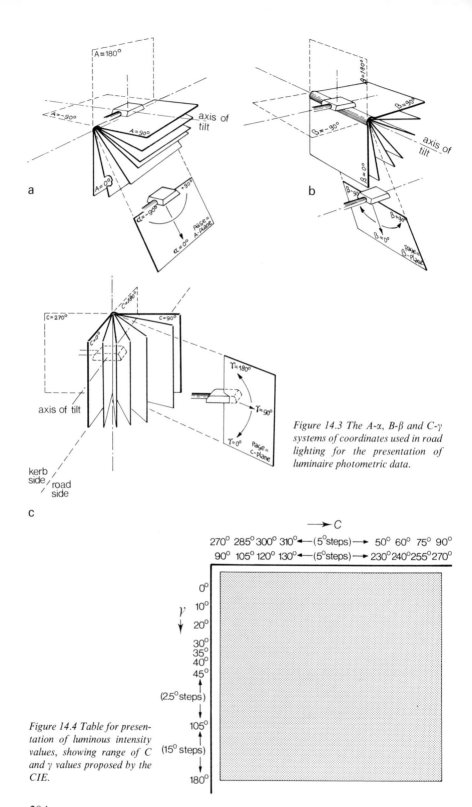

Figure 14.3 The A-α, B-β and C-γ systems of coordinates used in road lighting for the presentation of luminaire photometric data.

Figure 14.4 Table for presentation of luminous intensity values, showing range of C and γ values proposed by the CIE.

C/GAMMA I-TABEL

NE---> C

LE = γ	270.0	285.0	300.0	310.0	315.0	320.0	325.0	330.0	335.0	340.0
0.0	159.0	159.0	159.0	159.0	159.0	159.0	159.0	159.0	159.0	159.0
10.0	156.3	161.5	164.8	165.6	165.9	165.8	165.3	164.8	164.3	163.5
20.0	144.2	149.6	152.2	153.3	154.1	155.5	158.0	159.4	160.2	160.7
30.0	124.0	127.0	136.2	144.9	149.9	154.3	158.4	161.7	163.5	164.5
35.0	113.0	116.2	129.4	142.5	149.3	155.3	160.4	164.2	165.5	166.5
40.0	102.3	105.9	122.7	139.7	148.5	156.7	162.0	165.6	167.3	168.4
45.0	93.0	96.5	116.8	136.8	146.3	154.1	161.5	166.5	168.8	170.3
47.5	88.0	92.2	114.4	135.5	144.7	153.0	161.2	167.3	169.8	171.4
50.0	83.0	88.1	111.7	134.1	143.4	152.4	161.2	168.5	171.0	172.8
52.5	78.0	83.8	108.5	132.2	142.0	151.9	161.6	169.3	172.6	174.6
55.0	72.8	79.4	105.5	129.6	141.1	152.3	163.1	171.7	175.3	177.5
57.5	67.5	74.9	102.6	127.6	140.7	153.6	165.8	176.4	180.9	181.8
60.0	62.0	70.2	100.0	126.6	142.3	157.9	169.8	185.6	191.9	192.3
62.5	57.0	65.7	97.9	128.7	148.0	167.4	186.5	203.1	207.4	208.0
65.0	51.8	61.4	97.6	134.6	157.9	181.9	202.8	217.8	222.9	224.2
67.5	46.5	58.0	99.0	144.5	174.3	202.4	214.1	223.3	230.3	230.3
70.0	41.2	55.1	101.8	155.1	185.6	215.1	211.5	211.7	215.6	211.7
72.5	36.0	52.7	103.5	159.1	179.7	192.7	184.4	176.0	174.3	167.3
75.0	31.0	50.7	102.3	149.4	155.4	149.2	133.9	119.6	112.7	111.9
77.5	27.5	46.2	96.6	126.8	119.8	102.5	90.2	77.7	70.9	71.4
80.0	24.0	40.5	84.0	98.9	84.9	66.0	60.1	53.7	50.9	49.9
82.5	20.5	34.0	70.2	72.0	60.7	47.9	42.4	39.7	38.6	37.9
85.0	17.0	27.4	55.8	49.2	41.7	34.6	31.3	29.1	28.6	27.9
87.5	13.5	21.4	41.3	31.2	27.4	24.5	22.9	21.6	21.3	21.1
90.0	10.0	16.0	27.0	18.7	17.5	16.9	16.0	15.0	14.9	14.8
92.5	8.5	13.4	20.5	15.7	14.7	14.0	13.3	12.5	12.4	12.3
95.0	6.3	9.6	15.7	13.4	12.5	11.7	11.1	10.5	10.3	10.0
97.5	3.5	5.0	12.4	11.5	10.8	10.1	9.9	10.0	9.8	9.7
100.0	1.2	1.3	10.1	10.0	9.5	9.1	9.4	9.8	9.5	0.0
102.5	0.0	0.0	8.5	8.8	8.4	8.3	8.9	6.0	0.0	0.0
105.0	0.0	0.0	6.8	7.7	7.7	7.4	2.8	0.0	0.0	0.0
120.0	0.0	0.0	0.0	0.0	0.0	0.0	0.0	0.0	0.0	0.0
135.0	0.0	0.0	0.0	0.0	0.0	0.0	0.0	0.0	0.0	0.0
150.0	0.0	0.0	0.0	0.0	0.0	0.0	0.0	0.0	0.0	0.0
165.0	0.0	0.0	0.0	0.0	0.0	0.0	0.0	0.0	0.0	0.0
180.0	0.0	0.0	0.0	0.0	0.0	0.0	0.0	0.0	0.0	0.0

Figure 14.5 Part of a table of luminous intensity values, of the kind illustrated in figure 14.4, drawn up for a specific luminaire type.

for a given luminaire irrespective of the luminous flux of the lamp(s) contained in it – so long as the intensity distributions of the various bare lamps are similar – the tabulated values are normally quoted in terms of so many candela per 1000 bare-lamp lumens. This method of presentation has the additional advantage that any design change in the lumen output of a particular lamp type will not affect the validity of the luminous intensity table drawn up for the luminaire in which it is used.

14.1.2 Road Surface Reflection

It was shown in Chapter 8 that the reflection properties of a road surface can be conveniently presented in the form of a reflection table containing reduced luminance coefficients.

Such a table is drawn up by performing a large series of laboratory measurements on a suitable sample of the road surface in question. The sample being measured must, of course, be as representative as possible of the road surface as a whole. This calls for very careful visual inspection prior to cutting-out as well as great care in handling afterwards. The cutting-out itself must be performed in such a way that the sample is not distorted or its surface damaged. To guard against distortion after cutting-out, the sample should immediately be supported in some way – for example, in a cement mould. The measurements should then be completed as soon as possible, before significant surface changes can take place due to drying out.

The size of the measuring area permitted by the sample is also important: too small, and the chances are that it will not be truly representative of the surface from which it was removed; too large, and the measurement will not be a 'point' measurement. A typical sample size for laboratory measurements would be somewhere between 5 cm × 20 cm and 10 cm × 30 cm.

The aim of the measurements is to determine the r-values of the sample for a specific set of combinations of β and $\tan\gamma$. (β and $\tan\gamma$ are defined in Chapter 8, see figure 8.2.) This set of combinations (specified by the CIE – CIE, 1976a) is indicated in figure 14.6.

In fact what is measured is not the reduced luminance coefficient (r) but the luminance coefficient (q), from which r can be derived using

$$r = q \cos^3\gamma$$

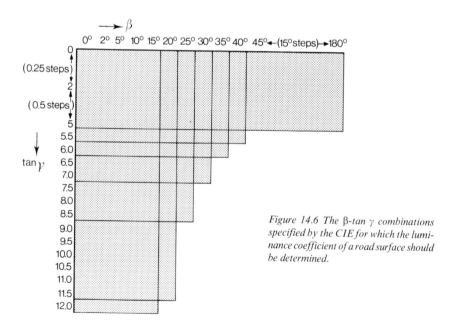

Figure 14.6 The β-tan γ combinations specified by the CIE for which the luminance coefficient of a road surface should be determined.

Since q is defined as the ratio of the luminance to the illuminance (Chapter 8), the q values for each combination of β and γ can be obtained from a series of luminance and illuminance measurements.

In practice it is possible to calibrate the measuring set-up in such a way that only luminances need be measured. Figure 14.7 illustrates the principle behind the two measuring set-ups commonly employed. The range of angles from $\beta = 0°$ (meter facing the light source) to $\beta = 180°$ (meter looking away from source) are set by rotating the road sample together with the meter about the vertical axis through the point of measurement.

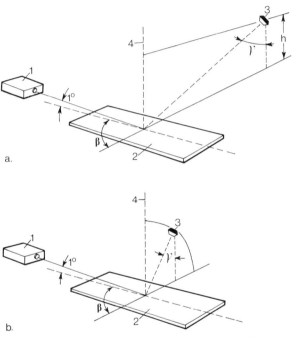

Figure 14.7 Two measuring set-ups used in determining the luminance coefficient of a sample of road surface (see text): 1. Luminance meter 2. Sample 3. Light source 4. Axis about which sample is rotated.

The γ angle can be set by moving the light source relative to the sample in either of two ways. In the technique illustrated in figure 14.7a, the source (always directed towards the point of measurement) is moved along a straight line such that its vertical distance above the sample remains constant. The solid angle subtended by the source at the point of measurement will, in this case, increase with decrease in the angle of light incidence (smaller angle γ); as is the case in an actual road lighting installation.

A practical measuring set-up of the constant-source-height type is shown in figure 14.8 (Verbeek and Vermeulen, 1971).

Alternatively, as illustrated in figure 14.7b, the light source can be moved over an arc about the sample. Here the solid angle subtended by the light source at the point of measurement is constant, although it is possible to introduce a variable-aperture diaphragm in front of the source to make the necessary correction to its size for change in angle γ (Ziegler, 1978). The advantage of moving the source about the sample in this way rather than away from it is, of

Figure 14.8 A practical measuring set-up based on the principle illustrated in figure 14.7a. The wheeled console contains automatic control and recording equipment.

course, that it permits the overall size of the measuring set-up to be reduced considerably.

There is in fact a third technique whereby the angle of light incidence can be easily varied, and that is to employ not one moveable source, but a number of fixed sources located so as to give the necessary range of γ angles and switched on in turn (photoflash discharge lamps have been used for this purpose – Von Berg, 1978). Such a set-up has the advantage that the time needed to complete the rather large series of measurements (an r-table contains 396 measured values) can be kept to a minimum, and this is especially important where wet-sample measurements are concerned if inaccuracies due to premature drying are to be avoided.

14.2 Field Measurements

14.2.1 Preliminary Considerations

Before taking a look at some of the instruments and techniques employed when making field measurements, it is necessary to say a few words concerning field measurements in general. These preliminary remarks can be placed under four headings: type of measurement; measuring points; observation points; and external influences.

Type of measurement

Generally speaking, where the aim is to determine the overall quality of an installation this has to be done by way of luminance measurements – luminance being the photometric basis of road lighting quality – combined with glare measurements.

Where, however, the aim is to check whether or not a luminance-designed installation is in compliance with specification, this can be done on the basis of illuminance measurements alone. If a check of the illuminance pattern on the road surface reveals that the measured values are in close agreement with those calculated, then the luminance pattern that will result will also be in accordance with that designed. This is provided, of course, that the road surface reflection properties are close to those on which the design was based. Such an illuminance check is the only approach in the case of an installation for a stretch of newly-laid road, the lighting design for which has been based not on the relatively unstable reflection properties of the new surface but on the reflection properties expected to exist after a specified 'running-in' period (see Sec. 8.2). The illuminance check can, unlike a check on luminance, be safely made during the running-in period.

The fact that it is often possible to fall back on an illuminance survey means, of course, that it is advisable to have available a complete set of calculated illuminance values for the installation being investigated, even if its design is entirely luminance based.

Knowing the designed illuminance values, it is also a simple matter to check whether an installation is in need of maintenance, modification, or even replacement. The measurement involved here is again one of illuminance, any deterioration in the quality of the lighting being made immediately apparent by comparing the calculated with the measured illuminances.

A quite different kind of field measurement dealt with here is that of assessing the reflection properties of a road surface actually on the road itself.

Measuring points

Whatever the type of measurement to be made, a decision must be reached concerning the number and relative positions of the measuring points: in other words, the measuring grid must be defined.

When all that is needed are a few spot checks to see that the measured values are in agreement with those specified in the design, just a few measuring points will suffice – although these must, of course, be evenly distributed over the road surface in question. If, in addition to comparing the measured values with the design values, these same points are used to give a rough indication of the measured average lighting level, then each measured value should be weighted in accordance with its position on the measuring grid.

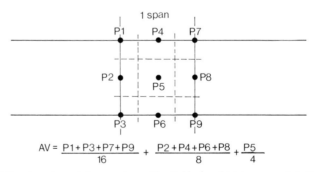

$$AV = \frac{P1+P3+P7+P9}{16} + \frac{P2+P4+P6+P8}{8} + \frac{P5}{4}$$

Figure 14.9 Regular, nine-point measuring grid suitable for obtaining a rough indication of the measured lighting level (in terms of either E_{av} or L_{av}) in which values are weighted according to the formula shown.

Figure 14.9 illustrates the layout of a nine-point measuring grid of the kind sometimes used when checking new road lighting installations. Explained in the figure is the weighting procedure that should be followed when calculating the average lighting level.

Where extreme accuracy is required, the CIE recommends using as many measuring points as specified earlier in Sec. 13.1.1, figure 13.1.

Observation points

Luminance, unlike illuminance, cannot be specified without specifying also the direction fróm which it was measured. When performing luminance measurements in connection with road lighting it is therefore necessary to define the so-called observation point or points employed – points defining the positions of the luminance meter relative to the points whose luminance is being measured.

For most purposes a few spot checks on the luminance will suffice and only one observation point is needed, preferably at a quarter of the road width from the near-side kerb (the so-called Standard Lateral Observer Position). For determining very accurately the quality of an already existing installation, the recommendations of the CIE regarding the position of the observation points as outlined earlier in Sec. 13.1.2 should be followed.

External influences

There are three factors that commonly exert an influence on the results obtained by measurements: the presence of stray light from light sources not belonging to the installation under consideration; variations in the supply voltage to the lamps; and, in the case of luminance measurements, temporary deviations in the reflection properties of the road surface.

Stray light reaching the road surface direct from shop windows, traffic signs, vehicle lights, and even from the moon, as well as light reflected onto the surface from light-coloured clothes, walls of nearby buildings and so forth, can add considerably to the amount of light reaching the road surface from the lighting installation alone. The influence of stray light is, of course, particularly noticeable when seeking to measure minimum values. Suppose, for example, that the minimum illuminance given by the road lighting installation alone at a point on the road surface is 0.5 lux; stray light contributing only 0.05 lux will here already account for a 10 per cent discrepancy. But even stray light that does not appreciably influence the illuminance measurement (light from an oncoming car, for example) can, due to its direction of incidence, have a marked effect on the magnitude of the luminance reading. Ideally, therefore, all stray light should be prevented from reaching the road surface, although the screens employed for the purpose must be so positioned that no light from the installation itself is lost.

For measurements to be meaningful, the installation's supply voltage should be kept to its nominal value while the measurements are being made; only then will the nominal flux from the lamps be obtained. If control of the supply voltage is not possible, it should at least be monitored so that in the event of supply variations occurring, corresponding corrections to the light output can be made. For installations containing new discharge lamps, 100 hours of operation should have elapsed before measurements are taken, in order to be sure that the lamps are stabilised.

Before luminance measurements are carried out it is important to check that the normal reflection properties of the road surface are not temporarily changed for one reason or another. Such changes can be caused by large accumulations of surface dirt (mud, oil, leaves, salt – employed to melt snow and ice – and so forth) or, in the case of bitumen-based surfaces during very hot weather, by partial melting or softening of the surface itself.

Where it is the dry-surface luminance that is being measured, care must of course be taken to ensure that the surface is perfectly dry. (An additional source of error or uncertainty will arise where the wet-surface luminance is being measured, since this surface condition is not accurately reproducible.)

Finally, since field measurements are only valid for the conditions prevailing at the time of the survey, it is important to record a detailed description of the area surveyed and to make a note of all those factors that might conceivably influence the results; factors such as lamp type and age, luminaire and ballast type, supply voltage, state of maintenance (including date last cleaned), weather condition (temperature, humidity – see following section) exact location of the observation and measuring points and the type and serial number of the measuring instrument used.

14.2.2 Illuminance Measurements

The illuminance meter, or luxmeter, most suitable for the measurement of illuminance in the field uses a class of photocell called a photovoltaic or barrier-layer cell. This cell (usually either selenium or silicium) is able to convert the light incident upon it into electrical energy without the aid of an auxiliary voltage supply; all that is essentially needed to complete the instrument is a microammeter scaled in lux. Some luxmeters of this type have a built-in amplifier to maintain measurement accuracy when working with the low illuminances often occurring in the field.

Cell corrections

$V(\lambda)$ correction. The spectral response of a photocell differs from that of the eye as defined by the spectral sensitivity $V(\lambda)$ of the CIE Standard Observer, on which all light units are based. For this reason, a correction filter or filters must be placed over the cell window – so-called $V(\lambda)$ correction filters. The various makes of cell, and even cells from the same manufacturer, differ in the degree of correction needed to match cell to eye. Each cell should therefore be supplied with the appropriate correction filter. It should be realised, however, that exact correction by means of filters is very difficult to achieve, and where the correction is less than perfect a change in the spectrum of the light being

212

measured will call for the application of a correction factor to the meter readings. The correction factors (or colour correction factors) for different light sources can be obtained by having the luxmeter calibrated.

Unless the luxmeter is equipped with an amplifier, it may be necessary to remove the light-absorbing correction filter in order to obtain sufficient sensitivity for measurement at low lighting levels. Where this is the case, an additional calibration to determine the correction factor needed under these conditions is called for.

Since photocells undergo an ageing process which affects both their sensitivity and their spectral response, it will be necessary to have the luxmeter recalibrated at least once a year.

Cosine correction. Another possible source of measurement error, if not corrected for, can occur due to the fact that at large angles of light incidence a large proportion of the light is reflected by the flat surface of the cell and therefore does not contribute to its photoelectric output.

To obtain the correct cell response under such conditions special devices called cosine correctors should be used. The cosine corrector consists of a semi-translucent glass or plastics disc or dome which is placed over the cell window in a specially shaped holder. For measurements in road lighting, the photocell should be corrected for angles of incidence of up to 85°.

Needless to say, a cosine corrector cannot compensate for variations in the alignment of the instrument itself, and for this reason it is preferable to perform measurements of horizontal illuminance with the photocell cardan, or gimbal, mounted so that it is always in the true horizontal.

Moisture on the photocell, or on any filter or corrector in use, can detract considerably from the accuracy of the measurements. The cell should therefore be kept dry, and it is better to avoid making measurements altogether during periods of extremely high humidity.

Some photocells are affected more than others by changes in temperature. Where this affects measurement accuracy it will be necessary to have the instrument calibrated so that appropriate corrections can be made to the readings obtained.

Finally, it should be appreciated that the accuracy obtainable in the field when measuring illuminance is normally no better than about ± 10 per cent, even when all the precautions detailed above have been put into effect (Walthert, 1977a and Van Bommel, 1978a).

14.2.3 Luminance

The instrument employed to measure the luminance at a 'point' on a road surface must have a high measurement sensitivity, a small measuring field and be equipped with an accurate aiming device. The small measuring field and the ability to aim the instrument accurately at the very small area being measured are necessary in order that the instrument should 'see' only this small area and no other. (At a distance of 90 metres, for example, a lengthwise stretch of road 3 metres long viewed from a position 1.5 metres above the road surface – the standard observer height used in road lighting – will subtend a vertical angle at the eye of just 2 minutes of arc.)

The photocell employed in luminance meters suitable for making measurements in road lighting is usually the photomultiplier. As is the case with the photocells of luxmeters, the photomultiplier tube of a luminance meter has to

Figure 14.10 A luminance meter for use in the field: (a) The instrument tripod mounted and ready for use.

214

be colour corrected by means of filters to match its spectral response to the spectral luminous efficiency curve – $V(\lambda)$ correction. Again, because this correction is seldom exact, it is sometimes necessary to apply colour correction factors to the meter readings, according to the spectrum of the light being measured.

The accuracy achievable in the field when measuring point luminance is normally no better than about 20 per cent, even under the most favourable circumstances (Walthert, 1977a and Van Bommel, 1978a).

A typical luminance meter

The luminance meter described here is typical of those currently being used in the field for making direct measurements of road surface luminance. Developed by Morass and Rendl (1967), this very sensitive instrument (figure 14.10a) can be used both as an integrating luminance meter for measuring average luminances as well as a spot meter for measuring luminance distribution.

The instrument is shown diagrammatically in figure 14.10b. An image of the area to be measured is formed by the objective on the measuring mask. The mask consists of a silvered, reflecting surface in which is a partly reflecting

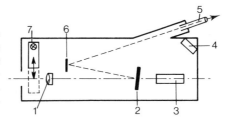

(b) Schematic cross-section showing the position of the main components referred to in the text: 1. Objective lens 2. Measuring mask 3. Photomultiplier 4. Meter 5. Eyepiece 6. Mirror 7. Calibrating light source.

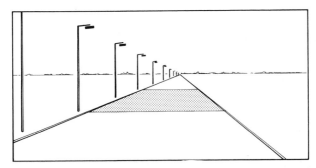

(c) View obtained through the instrument of the road being studied, showing the dark patch of the measured area. Different trapezium-shaped measuring masks are needed for different road widths and viewing positions.

glass aperture. Light passing through the aperture is incident upon the photo-multiplier, the output of which gives a reading of luminance on the meter. At the same time, light reflected by the glass of the aperture and by the remainder of the mask is passed via a mirror to the instrument's eyepiece. The area being measured and its surroundings are thus simultaneously visible, the former appearing as a dark patch on the latter, figure 14.10c.

A range of measuring masks, each with a different-sized aperture, allows the size of the measuring field to be selected – a very small aperture for spot measurements and trapezium-shaped apertures for measuring the average luminance of straight stretches of road. (Strictly speaking, real spot measurements cannot be made even with the smallest aperture conceivable. However, measurements made with apertures giving a field of view equal to or smaller than 2 minutes of arc vertically by 20 minutes of arc horizontally can in practice be treated as being point measurements (CIE, 1976a).)

The instrument is provided with a built-in calibration standard, which can be moved across in front of the objective. Also built-in is a $V(\lambda)$ correction filter. When the meter is used for the measurement of average road surface luminance, the perspective average is obtained (See Sec. 13.1.3). In order to determine the dynamic average luminance (See again Sec. 13.1.3), two measurements with the meter should be made, one with the measuring field beginning under a luminaire and one with it beginning midway between two adjacent luminaires. The average of these two measurements will be a good approximation of the dynamic average luminance which, in turn, corresponds to the arithmetic average of point luminance over a regular grid, as explained in the section referred to above.

A luminance recording system

A comparatively recent development in road lighting is the use of closed-circuit television systems to automatically measure, display and record the wealth of luminance information needed when conducting a full-scale lighting survey.

One such system (Balder and Meulders, 1969) operates as follows. A television camera views the road from the normal driving position of a road user and produces a picture of what it sees on the screen of a closed-circuit television monitor. The television picture is made up in the normal way, viz. by a process of line scanning. However, whilst each picture line is being scanned (there are 625 lines to a complete picture), the fluctuating amplitude of the luminance signal (the signal that alters the brightness of the spot on the screen in response to differences in luminance occurring along horizontal picture elements) is examined by a special circuit. Each time this circuit detects that the amplitude of the luminance signal is exactly equal to a specific value (which can be preset) an extra pulse is added to the signal, which causes an extra bright spot to appear on the screen of the monitor. Thus, the appearance of bright spots in the television picture means that at these spots on the road the luminances are equal, a line of such spots forming a line of isoluminance (figure 14.11). By adjusting the preset comparison level of the luminance signal to different values, a series of isoluminance lines can be generated, so revealing the entire luminance pattern of the road surface being viewed.

Figure 14.11 The picture on the television monitor when the camera is installed in the mobile laboratory or 'light van' of figure 14.12. Digital readouts show, from left to right: L_{av}, U_l and L_v. The vertical white bar, centre, indicates the line along which U_l is measured, while the curving line of light points on the road surface form an isoluminance contour.

Longitudinal uniformity is measured by another circuit which examines the amplitude of the luminance signal as the scan crosses a vertical bar displayed on the screen. At the completion of each picture scan, the minimum and maximum luminance values detected within the bar are electronically divided and the value of the ratio L_{min}/L_{max}, the longitudinal uniformity ratio, displayed on the screen (see again figure 14.1L). By manoeuvring the TV camera so that the bar occupies the desired position on the picture, the longitudinal uniformity at various distances from the kerb can be measured.

Colour plate 2 (facing page 97) shows an open-air lighting laboratory where a system similar to the one described above is installed. The laboratory enables various full-scale lighting designs and items of lighting equipment to be tested with the minimum of time spent on preparation.

A mobile laboratory (Van Bommel, 1976a) equipped with a luminance recording system is shown in the photograph of figure 14.12. Apart from displaying

Figure 14.12 A mobile road lighting laboratory. The trailer contains a selection of luminaires, any of which can be mounted atop the telescopic lighting column. Power comes from a generator housed in the same trailer.

Figure 14.13 The cab of the mobile road lighting laboratory showing: left, the luminance meter; centre, a glare meter; top, the television camera; and bottom left, the television monitor.

the results of all measurements for immediate use, all data fed to the monitor are here automatically recorded on video-tape for detailed analysis at a later date. This means, of course, that measurements can be made while the mobile laboratory is on the move, so reducing the time needed to make a full-scale lighting survey from days, of even weeks, to a few hours.

Amongst the other measuring equipment installed in this vehicle are an integrating luminance meter and a glare meter (figure 14.13) – see description below.

14.2.4 Glare

The common measure of disability glare is the threshold increment, *TI*, given in per cent by

$$TI = 65 \frac{L_{\mathrm{v}}}{L_{\mathrm{av}}^{0.8}}$$

where L_v is the total equivalent veiling luminance in the situation considered and L_{av} the average road surface luminance.

The total equivalent veiling luminance can be measured with the aid of a luminance meter fitted with a glare lens. The glare lens, placed in front of the luminance meter, weights the light passing through it in such a way that the L_v of the scene being viewed is measured direct with just a single reading. (Fry, *et al.*, 1963, describe how such a lens can be designed and calibrated; this design is based on a formula for the veiling luminance that is slightly different to the one given in Sec. 1.2.3.) The average road surface luminance is, of course, measured in the normal manner, without the glare lens attached.

For discomfort glare, there is the glare control mark (*G*). However, due to the psychological nature of discomfort glare, the value of *G* cannot be derived from measurements, as can *TI*, but must be arrived at by computations based on the photometric data of the luminaire and of the installation in which it is being used. These computations are to be found in Chapter 13 Calculations.

14.2.5 Road Surface Reflection

Portable reflectometers suitable for the measurement of *S*1 and Q_o (and possibly *S*2, the description parameters of a road surface – Sec. 8.1.1) in the field are not yet generally available. There are, however, a number of prototype field instruments in use and it is proposed here to outline briefly the various principles of operation employed in these.

The prototypes can be broadly divided into two groups according to whether Q_o is measured direct or must be calculated from the results of other luminance measurements. For instruments of both groups, *S*1 is obtained by taking the ratio of two measured luminances.

Direct reflectometers

Q_o *measurement*. The average luminance coefficient at a point on a road surface is directly proportional to the luminance at that point when that point is lighted from positions of equal height above the road surface evenly distributed within the Q_o integration boundaries (defined in figure 8.5) and provided that the light is incident upon the point from all directions with equal luminous intensity (De Boer and Vermeulen, 1967). All direct reflectometers make use of this principle.

The constant luminous intensity is provided in practice by a flat luminous 'ceiling' above the point of measurement. This ceiling consists of a diffusely scattering filter lit from above by a regular array of lamps. Such a filter is shown in figure 14.14, while figure 14.15 shows the filter in position in the reflectometer. The luminance meter built-into the instrument views the point illuminated by the luminous ceiling via a small mirror which is so angled as to

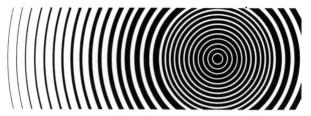

Figure 14.14 Compensation filter employed in the reflectometer of figure 14.15 to create the correct luminous 'ceiling' above the point of measurement when determining Q_0.

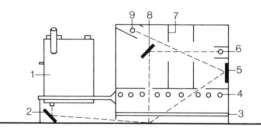

Figure 14.15 Schematic cross-section of a direct reflectometer showing: 1. Luminance meter 2. Mirror 3. Compensating filter 4. Lamps used in measuring Q_0 5. Mirror 6. Lamp used in measuring r (0,0) 7. Baffle 8. Mirror 9. Lamp used in measuring r (0,2).

produce the required $1°$ standard observation angle with the horizontal (usually a system of mirrors is used for this purpose).

S1 measurement. The specular factor *S1* is defined as the ratio of the two reduced luminance coefficients r ($\beta = 0°$, $\tan\gamma = 2$) and $r(\beta = 0°, \tan\gamma = 0)$ – normally written $r(0,2)$ and $r(0,0)$ for short.

Looking again at figure 14.15 it can be seen that there are two small lamps situated in the upper part of the instrument to give the required angles of light incidence. With the filter removed and these lamps switched on in turn, the two reduced luminance coefficients can be measured using the instrument's luminance meter. (It is also possible to calculate *S2* once the $r(0,0)$ and Q_0 values are known: $S2 = Q_0/r(0,0)$.)

Because of the need to maintain a high degree of accuracy in the alignment of the instrument's various parts (mirrors, filter, baffles, etc.) portable reflectometers must be of very sturdy construction. For this reason, they tend to be rather bulky and heavy also. Figure 14.16 shows a reflectometer of the type described above in position on the road. As can be seen, it is fitted with wheels and a towing handle to facilitate movement. Calibration of the instrument is performed with the aid of a sample having known Q_0 and *S1* values. Checking the alignment of the various parts may have to be done fairly often, especially if the instrument is used on rough roads.

Figure 14.16 The reflectometer of figure 14.15 in position on the road, with compensating filter about to be inserted. The area measured under the 50 cm × 20 cm filter is 8 cm × 0.8 cm.

Indirect reflectometers

Q_0 *measurement.* Reflectometers intended for the determination of Q_0 via the measurement of related luminance quantities are based on the correlation known to exist between Q_0 and certain reduced luminance coefficients, or r-values: by measuring the latter, the former can be calculated.

One such reflectometer (Range, 1973), which is based on the correlation existing between $Q_0/r(0,0)$ and the two reduced luminance coefficients $r(0,1)$ and $r(180,0.375)$, is shown in figure 14.17. The instrument is both portable and easy to use, the lamp being simply moved along the arc to give the three angles of light incidence at which measurements have to be made: $\gamma = 0$ for $r(0,0)$; $\gamma = \tan^{-1}1$ for $r(0,1)$; and $\gamma = \tan^{-1}0.375$ for $r(180,0.375)$.

Another instrument (Øbro and Sørensen, 1976) measures eight different reduced luminance coefficients using eight lamps which, fixed in position, are switched on in turn to illuminate the measuring field. The value of Q_0 is then calculated by means of a linear equation.

A very reliable correlation of the type described above is that recently found to exist between $Q_0/r(0,0)$ and the reduced luminance coefficient $r(5,5)$ – Burghout, 1977. This correlation can, in principle, be used when determining Q_0 via the measurement of $r(0,0)$ and $r(5,5)$. Unfortunately the measurement of $r(5,5)$ involves a rather large angle of light incidence, which makes the measurement difficult, and so an instrument making use of this correlation does not as yet exist.

Finally, it is perhaps of interest to note that there is a further correlation that allows Q_o to be calculated from a measurement of $r(0,0)$ and the $S1$ value. The correlation referred to is that existing between $S1$ and the other specular factor, $S2$, shown earlier in figure 8.10. From this correlation, and given the fact that $Q_o = S2 \cdot r(0,0)$, it is a simple matter to calculate Q_o using the derived $S2$ value and the $r(0,0)$ value obtained whilst measuring $S1$.

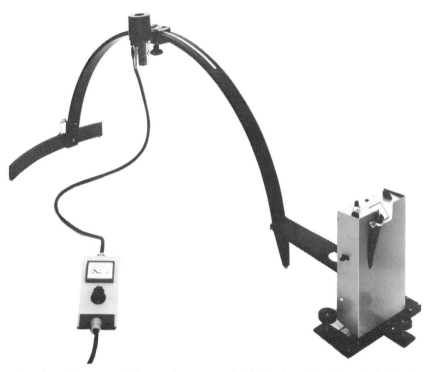

Figure 14.17 A compact, indirect reflectometer suitable for determining Q_o and $S1$. The instrument consists of a luminance meter, a lamp which can be moved over an arc about the point of measurement, and an ammeter for adjusting the lamp current obtained from an accumulator. The measuring area is 11.5 cm × 4.5 cm (Range)

S1 measurement. All reflectometers intended for use in determining $S1$ work on the principle outlined above in connection with the direct-reading reflectometer: viz. they all incorporate a means whereby the two angles of light incidence needed to measure $r(0,2)$ and $r(0,0)$ may be provided. They differ only in the way this principle is put into effect. Range's instrument, for example, employs a single moveable lamp, while that designed by Obro and Sørensen has two lamps which are fixed in position. Alternatively, a fixed lamp and a mirror system could equally well be employed to achieve the same effect.

222

Motorway interchange, Munich, West Germany with high-mast asymmetrical floodlighting from high-pressure sodium luminaires. (See Sec. 15.1.3)

Facade mounted luminaires in the style of the old gas lanterns, but housing modern 35 watt high-pressure sodium lamps dispense with the need for intrusive lighting columns and so help to preserve the old-world charm of this village street. (See Sec. 16.2.3)

7 *Motorway guidance lighting near Zwolle, the Netherlands. In the absence of a difference in source colour (top) the exit to the right is not readily apparent, especially from a distance. But in reality (bottom) the exit is lighted with high-pressure mercury which contrasts well with the low-pressure sodium used for the motorway and the motorway entrance. (See Sec. 15.2.1)*

Part 4

APPLICATION FIELDS

Chapter 15

Roads and Junctions

The fundamental lighting criteria that have to be considered when designing a road lighting installation are discussed in detail in Part 3 of this book.

In this chapter the typical characteristics of the various types of installation met with in practice will be reviewed, and this will be done in such a way as to facilitate the choice of installation best suited for each of a number of specific situations.

Since there is often a certain interaction between differing but adjacent installations, the harmonic integration of these will also be discussed.

15.1 Lighting Systems

15.1.1 Conventional Lighting

A conventional road lighting installation is one in which the luminaires are orientated such that their beam axes are pointing along or near to the road axis and where the spacing between adjacent luminaires is more or less constant.

The geometric variables in such an installation are the luminaire (or column) arrangement, mounting height, spacing, overhang and inclination.

Basic arrangements
There are five basic conventional arrangements, figure 15.1.

Single-sided. In a single-sided arrangement all the luminaires are located at one side of the road. With this type of arrangement the luminance of the road surface at the side farthest from the luminaires is usually lower than that of the side nearest to them (figure 15.2). In order, therefore, to ensure that the overall uniformity will be adequate, the single-sided arrangement should only be adopted in conjunction with mounting heights approximately equal to or greater than the effective width of the road. (The effective width is the horizontal distance between luminaire and far kerb, figure 15.3.) (The influence of mounting height on total costs of a single-sided installation is dealt with in Sec. 12.1.2.) Unlike the other arrangements dealt with here, the single-sided arrangement gives different lighting conditions for the two directions of traffic flow.

Visual guidance is good with this arrangement.

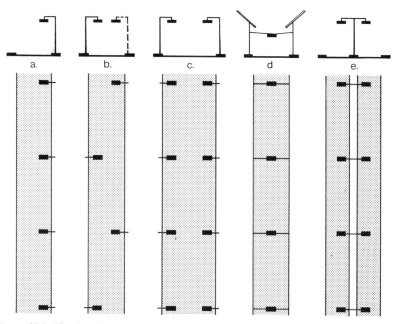

Figure 15.1 The five basic conventional road lighting arrangements: (a) Single-sided (b) Staggered (c) Opposite (d) Spanwire (e) Twin central.

Figure 15.2 A typical single-sided installation with mounting height equal to effective road width.

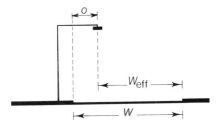

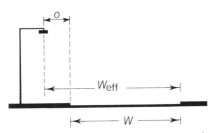

Figure 15.3 Effective road width (W_{eff}) equals actual width(W) minus luminaire overhang(O).

Figure 15.4 A typical staggered installation where mounting height equals two-thirds the effective road width. The leopard-skin effect mentioned in the text is clearly visible.

Staggered. Here the luminaires are placed alternately on either side of the road in a so-called zig-zag fashion. The overall uniformity obtainable by siting the luminaires in this way will be adequate so long as the luminaire mounting height is made equal to at least 2/3 of the effective road width. The longitudinal luminance uniformity is generally low, the alternate bright and dark patches creating a sort of 'leopard-skin' effect (figure 15.4). The zig-zag positioning of the luminaires can also sometimes create a rather confused impression of the run of the road. For these reasons, it may sometimes be advisable to employ instead a single-sided arrangement with increased mounting height, the higher cost of the taller columns perhaps being offset by the savings in trench cutting and cabling (one run of cable instead of two) that such an arrangement brings with it. It should be remembered, however, that the staggered arrangement does exhibit certain advantages over the single-sided one during periods of wet weather (Sec. 10.1.3).

Opposite. This arrangement, with the luminaires placed opposite one another (figure 15.5), is used mainly when the mounting height is less than 2/3 the effective road width, although mounting heights down to 2/5 of this width are in fact permissible. (The influence of mounting height on the total cost of an opposite installation is considered in Sec. 12.1.2.)

229

Figure 15.5 The opposite (or opposed) lighting arrangement with luminaire mounting height equal to one-half the effective width of the road.

When an opposite arrangement is employed on a dual carriageway having a central reserve wider than about 1/3 the carriageway width, it effectively becomes two independent single-sided arrangements and must be treated as such. The same, of course, holds true when obstructions (e.g. trees, anti-glare screens) are placed along the central reserve.

Spanwire. In the spanwire arrangement the luminaires are hung from cables strung across the road, the luminaires having their main vertical plane of symmetry at right angles to the road axis such that most of their light is thrown along the road rather than across it. (This arrangement should not be confused with the catenary arrangement, which will be dealt with later.)

The spanwire arrangement is normally used at rather low mounting heights (around 6 to 8 metres) for narrow roads in built-up areas when there is no possibility of employing road-side lighting columns. The suspension cables are then simply strung between buildings on either side of the road.

This arrangement may also offer a solution to the problem of where to site the luminaires such that their light output will not be shielded by roadside trees (figure 15.6).

The principle of using cables instead of columns to support the luminaires can, of course, be applied to any of the three arrangements considered above;

230

Figure 15.6 Spanwire lighting. Conventional road-side lighting columns would have been shielded by the dense foliage, while the tram lines clearly prohibit the use of any central arrangement.

the basic considerations governing the choice of arrangement will remain the same.

Twin Central. The twin central arrangement is intended for dual carriageways, the luminaires being mounted on T-shaped masts in the middle of the central reserve (figure 15.7). Since this is essentially two single-sided arrangements placed back to back, the condition that the luminaire mounting height should be equal to or greater than the effective road width again applies.

The luminance of the road surface in the lanes farthest from the luminaires, viz. the 'slow' lanes, will usually be lower than that on the 'fast' lanes.

With the twin-central arrangement both the road and kerb-side parts of the luminaire light output contribute to the luminance on the road surface, provided that is that the central reserve is not too wide. This is in contrast to the opposite arrangement, where it is mainly the road-side part of the luminaire light output that provides the luminance. The twin-central arrangement is therefore the more efficient of the two. (The total costs of the two types of installation are compared for various mounting heights in Sec. 12.1.2.)

231

Figure 15.7 A typical twin central lighting arrangement in which the mounting height is equal to the effective width of the road.

It should be noted, however, that the opposite arrangement may give slightly better lighting under wet-weather conditions; provided, that is, that the central reserve is rather narrow and does not contain obstructions (See Sec. 10.1.3).

The visual guidance obtained with the central arrangement is good.

Combined arrangements

Combinations of these five basic arrangements are, of course, also employed. The twin-central and opposite arrangements, for example, are often combined on dual carriageways (figure 15.8) to give in fact either a staggered or an opposite arrangement for the individual carriageways.

In order to obtain a specified lighting level from a given type of light source it is sometimes necessary to employ more than the usual one or two luminaires per mast (figure 15.9, and colourplate 3 facing page 128). The rules governing mounting height and effective road width given above for the various arrangements remain unchanged for these multi-luminaire installations.

Mounting height, spacing, overhang and inclination

Mounting height. Luminaire mounting height, dependent in the first place on the lighting arrangement and effective road width, as outlined above, has an

232

Figure 15.8 Twin central and opposite arrangements combined for the lighting of a motor-way.

Figure 15.9 Arabian Gulf Street, Kuwait. Each sixteen-metre-high lighting column carries six 400 watt high-pressure mercury luminaires to give an average road surface luminance of 2 cd/m².

important bearing on the cost of the installation and the degree to which its maintenance will be facilitated.

Often, the lower the mounting height the lower the overall cost of the installation (Sec. 12.1.2). But it should be borne in mind that for all luminaires glare (both disability and discomfort glare) increases with reduction in mounting height.

Spacing. Luminaire or column spacing for a given lighting arrangement and luminaire light distribution is dependent on the mounting height and the longitudinal uniformity planned for the installation. The greater the mounting height, the larger can be the spacing for a given longitudinal uniformity. One advantage of maximising on the spacing is that for a given longitudinal uniformity and driving speed the rate of change of luminance is minimised, with consequent improvements in driver reaction performance (Sec. 2.5) and driver comfort (Sec. 3.2.2).

Overhang. The amount of luminaire overhang (viz. the distance that the luminaire projects out from or is placed back from the kerb, figure 15.3) serves to determine the effective width of the road and thereby the minimum mounting height required for the luminaires. Care should, of course, be taken that – perhaps in an endeavour to improve the lighting quality in wet weather (See Sec. 10.1.3) – the overhang is not made so great that the public footpath(s) receives insufficient light (See Chapter 9 Recommendations).

The overhang should if necessary be varied such that, particularly in the interests of providing good visual guidance, the luminaires appear to form a smooth line in the driver's field of view. This will be necessary, for example, where the stretch being lighted includes certain types of discontinuity in the kerb line. Consider the case where the need arises to place a column or columns along a section where the road has been widened to provide room for a lay-by or bus stop (figure 15.10). Here, the columns lying back from the normal kerb line will need a correspondingly greater overhang in order to bring all the luminaires in line.

Inclination (or tilt). Inclining or tilting the luminaires up from the horizontal is done to increase the road width covered with a given mounting height. But this measure is not very effective. If the effective road width is large compared with the mounting height, tilting the luminaires will increase the amount of light reaching the far side of the road but the luminance here will not be increased in proportion. This is because of the unfavourable angle of light incidence relative to the approaching motorist.

Tilting, especially at bends in the road, also increases the chances of glare being produced and makes it difficult to provide good visual guidance. In

234

view of the above, it is recommended that the angle of tilt with respect to the normal angle of mounting be limited to an absolute maximum of 10°, a top limit of 5° being preferable.

Whilst it is possible to come to a tentative choice of lighting arrangement, mounting height, spacing, overhang and inclination for a given situation by paying close attention to the points mentioned above, the choice can only be finalised once everything has been checked by calculation to see that all the various photometric quality requirements have been satisfied (See Chapter 13 Calculations).

Figure 15.10 Luminaire overhang and visual guidance. Failure to increase the overhang where the luminaires are set back from the normal kerb line (top) creates confusion. The heavily retouched photograph of the same situation (below) depicts the improvement that would be brought about by locally increasing the overhang.

Curves

Curves having a radius in the order of 1000 m or greater can be treated as straight roads so far as lighting is concerned. The rules governing the siting of the luminaires are then as outlined in the foregoing sections of this chapter. The siting of luminaires on curves of smaller radius, however, should be given extra attention since this can influence both the luminance pattern on the road surface and the quality of the visual guidance obtained.

The preferred column arrangement is, as before, largely dependent upon the mounting height chosen and the effective width of the road. Where the former is at least 2/3 the latter, the columns should be placed in a single-sided arrangement along the outside of the curve; luminaires along its outside contribute more to the luminance of the road surface than do luminaires along its inside – whilst light from the former is mostly reflected by the road surface towards the road user, that from the latter is reflected toward him principally by areas adjacent to the road (figure 15.11a and b).

Luminaires at the outside of a curve also mark the run of the road better than do luminaires at the inside. At the same time, however, care should be taken to avoid creating the situation depicted in figure 15.12a, where the single badly placed luminaire may suggest to an approaching driver that the road con-

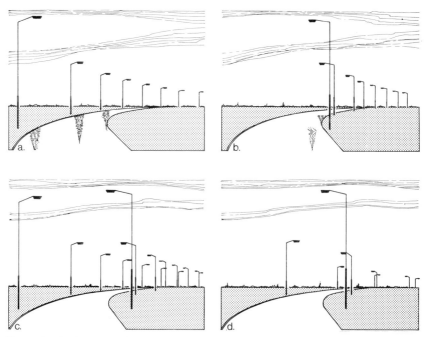

Figure 15.11 The preferred luminaire arrangement on a curve. Single-sided on outside (a) is preferable to single-sided on inside (b). Opposite arrangement (c) is preferable to staggered arrangement (d).

tinues straight farther than it really does, perhaps causing him to brake dangerously late. Figure 15.12b shows a better position for this luminaire. For smaller mounting heights (or wider roads) an opposite arrangement should be adopted, even though the visual guidance will not be so good (figure 15.11c). The staggered arrangement gives extremely poor visual guidance on curves (figure 15.11d) and should therefore be avoided.

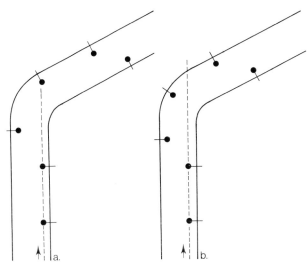

Figure 15.12 A potentially dangerous situation created through a badly-sited luminaire. (a) The second luminaire on the bend is in line with those on the straight and so may produce a misleading impression of the run of the road. (b) The offending luminaire correctly placed to avoid confusion.

The spacing of the luminaires should be closer for a curve than for a similar stretch of straight road, a reduction ratio of from 0.5 to 0.75 being quite common.

15.1.2 Catenary Lighting

In a catenary installation the luminaires are suspended from a cable (the catenary) stretched between widely-spaced columns placed along the central reserve between two carriageways (figure 15.13). The special catenary luminaires employed, which are orientated such that the longitudinal luminaire axis is parallel to the road, shed most of their light in a direction at right angles to this axis, that is to say across the road rather than along it. Because of this they must be spaced rather close together so as to give a spacing/mounting-height ratio of about 1.5. (Very occasionally, masts in-

Figure 15.13 A typical catenary lighting installation.

Figure 15.14 Catenary lighting is noted for its outstanding visual guidance and luminance uniformity.

stead of cables are employed to support the catenary luminaires, but the luminaire orientation is the same as with normal catenary lighting, as is the spacing/mounting-height ratio.)

Lighting characteristics
A catenary lighting installation combines good luminance uniformity with excellent visual guidance (figure 15.14).
Good overall uniformity (U_o better than 0.4) can in fact be obtained with a mounting height that is less than 2/3 the effective width of a carriageway: catenary luminaires have been developed that allow the mounting height to be reduced to even less than 1/2 this width (colour plate 4 facing page 129). A longitudinal uniformity (U_l) of 0.8 or more is quite common.
The deterioration in luminance uniformity that inevitably takes place when a road surface becomes wet is minimised with this type of lighting installation. Reduced also, is the disturbing light scatter that occurs in the presence of fog and mist (See Sec. 10.1.2 and 10.2).

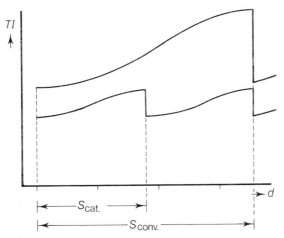

Figure 15.15 Periodic change in threshold increment (TI) – the glare stimulus – with distance travelled between successive luminaires for a conventional installation (top) and a catenary installation. Spacing catenary (S_{cat}) equals one-half spacing conventional (S_{conv}).

Because the luminaires of a catenary installation are viewed almost axially by a driver, the amount of discomfort and disability glare produced is small. Moreover, the fluctuation in the glare stimulus experienced by a driver by virtue of his movement along the road is less disturbing under catenary lighting (figure 15.15) than under any of the conventional lighting systems described above (Van den Bijllaardt *et al.*, 1975 and Westermann, 1975b).

The predominantly transverse light distribution possessed by the catenary luminaire means that the luminance yield is lower for catenary lighting than for conventional installations. Thus, for the same road surface luminance the horizontal illuminance given by a catenary installation must be greater than that given by conventional lighting. In practice, however, the quality of catenary lighting (i.e. its uniformity, visual guidance and low glare disturbance) is such that average luminance values equal to about 80 per cent of those normally recommended are often acceptable where catenary lighting installations are concerned, and this usually more than compensates for the lower luminance yield.

Although in principle an acceptable catenary luminaire can be constructed for most types of lamp, it is the low-pressure sodium lamp with its combination of ideal shape (long and slim) and relatively high luminous efficacy that is the favoured choice for catenary lighting. This is not to say that where improved colour rendering is required, the high-pressure sodium lamp will not give perfectly acceptable results.

Features of construction

The main features of construction of a typical catenary lighting installation are shown in figure 15.16. As can be seen, each luminaire hangs from the upper, or catenary, cable on a pair of short suspension cables. The luminaires are linked by a strainer cable which prevents them swaying out of position in a wind and also serves as a support for the electricity supply cable.

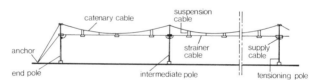

Figure 15.16 The main constructional features of a catenary lighting installation.

Most catenary systems are of the kind illustrated, in which three sorts of pole are employed. Individual lengths of catenary cable are fastened at each end to regularly-spaced tensioning poles (figures 15.17a). Between the tensioning poles the cable passes freely over the tops of a number of intermediate poles (figure 15.17b). At each end of the installation is an end pole or anchor pole, generally of heavier construction than the intervening poles and supported by stay wires, to which the catenary cable is terminated. Each strainer cable is individually terminated between poles.

This form of construction has a number of advantages over the alternative system in which the intermediate poles are replaced by tensioning poles. Erection is made both easier and cheaper as a whole section between two

Figure 15.17 Catenary masts: (a) A tensioning pole to which catenary and strainer cables are attached using tensioning devices. (B) An intermediate pole showing open slid at top, with lifting tackle in place ready to hoist catenary into position.

Figure 15.18 Assembling a section of catenary on the ground prior to it being hoisted into position.

tensioning poles can be assembled on the ground (figure 15.18). There is also less risk of a whole section of luminaires between two masts together with the electricity supply cable being dragged down to within reach of the ground in the event of a heavy collision with one of the intermediate poles.

Pole length is mainly dependent upon the mounting height required for the luminaires. For example, for a luminaire mounting height of 10 m the length of the poles will be about 16 m, 2 m of which is buried in the ground.

In present-day installations, the pole spacing can range from 60 m to 100 m. Pole strength is dependent upon pole height and pole spacing, the latter determining the number of luminaires between successive poles.

The strength of pole needed to support the system can be kept to a minimum by housing the lamp ballasts in the base of the pole, or else in special cabinets near to the poles, instead of in the luminaires themselves.

Needless to say, the factors influencing the cost of a catenary installation will, for a given country, be largely dependent upon the type of construction employed, the materials used and on whether or not any practical experience with these relatively new systems is available. The best that can be done here is to refer the reader to a cost evaluation that was performed in the Netherlands (Van den Bijllaardt *et al.*, 1975).

15.1.3 High-mast Lighting

The term high-mast lighting is generally used to describe lighting in which the luminaire mounting height is 20 m or more, the luminaires normally being mounted several to a mast to give the necessary degree of light coverage.

This form of lighting is often employed in preference to conventional road lighting for complex junctions on main roads, for motorway interchanges, and for large open spaces such as public car parks. The principal attraction of high-mast lighting in such applications is that it leaves the lighted area almost free of columns and so gives the road user an uncluttered view of the road junction and its exits.

Glare is also often less of a problem with this form of lighting, even when the junction involves a difference in height between roads, for the luminaires themselves can be placed virtually out of sight by careful siting of the lighting masts.

Maintenance can often be carried out without having to disturb the traffic flow.

Mounting arrangements

Three luminaire mounting arrangements are commonly in use: plane symmetric, radial symmetric, and asymmetric (figure 15.19).

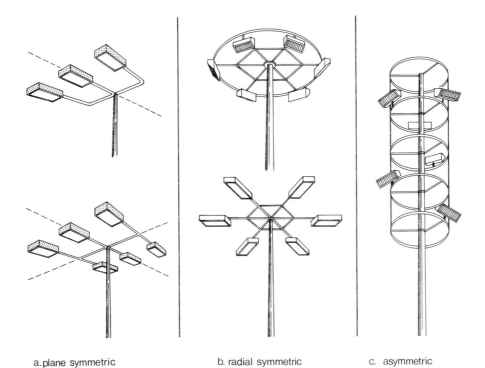

a. plane symmetric   b. radial symmetric   c. asymmetric

Figure 15.19 High mast mounting arrangements.

Figure 15.20 Three of the six thirty-three-metre-high masts used to light the Cherbourg interchange at La Défense, Paris. Each mast carries eight narrow-beam and eight wide-beam floodlight luminaires in a radial symmetric array.

243

Plane symmetric. The first of these arrangements is generally employed in conjunction with conventional luminaires and is particularly suitable for the lighting of more or less straight runs of wide road. Each luminaire is orientated at right angles to the road, with the major part of the light being thrown along the road as in a conventional installation, and the masts are spaced at regular intervals along it (See figure 15.9).

Because this sort of lighting has much in common with conventional lighting and must conform to the requirements governing the latter, it is sometimes described as 'semi high mast lighting', especially when intermediate mounting heights (say 16 to 18 metres) are employed.

Radial symmetric. Full symmetry in the way the luminaires are arranged on the mast is used where the aim is to spread the light more or less evenly in all directions. Such a distribution, which can be obtained using either conventional luminaires, or floodlights, is often suitable for lighting rather compact road junctions from a central position (figure 15.20). Large areas that need to be evenly illuminated with a minimum number of pole obstructions can also be tackled in this way (figure 15.21). The mast height to mast spacing ratio (adjacent masts) needed to ensure good uniformity is generally in the order of 1 : 3, depending upon the type of luminaire employed.

Figure 15.21 Car park lighting at Orly Airport, Paris. Sixteen thirty-metre-high lattice steel masts, each carrying thirty 400 watt high-pressure sodium floodlights in a radial symmetric array, have been used to light the 170000 m² of car park.

At motorway interchanges and similar grade-separated junctions there is the risk with this type of lighting that the elevated sections of the road relatively close under the luminaires will be unevenly illuminated. Whilst it is possible to minimise this effect by careful siting of the masts, it can seldom be avoided entirely as the choice of mast position is usually severely limited.

244

Radial symmetric high-mast lighting is sometimes chosen partly on aesthetic grounds (figure 15.22).

Figure 15.22 The Van Brienenoord motorway interchange, Rotterdam, the Netherlands. This radial symmetric high-mast installation consists of arrays of twenty-four low-pressure sodium floodlights mounted between twenty-eight and forty metres above the ground.

Asymmetric. In the third of the luminaire arrangements illustrated in figure 15.19 symmetry is totally lacking. Here each luminaire (floodlights only are used) is individually aimed so as to make the best possible use of the light available (figure 15.23 and colour plate 5 facing page 224). This is a technique especially well suited to the lighting of complex junctions, particularly those in which the carriageways are spread out over a rather large area, as it allows the light to be directed where it is needed, viz. on the road and not on its surrounds.

With this form of high-mast lighting the spacing between adjacent masts may be as much as 4 times the mast height. Glare can be avoided in critical directions by fitting those floodlights concerned with special screens or louvres (figure 15.24).

Because of the directional nature of floodlighting, measures should always be taken to ensure that lamp failure cannot result in a 'gap' in the lighting. With twin-lamp floodlights this problem does not of course arise, but where single-lamp units are in use it is best to have at least two of these aimed at each point on the junction where continuity of lighting is critical.

Cost

Whilst it is true that most high-mast lighting installations automatically provide a clearer view of a road junction than do conventional low mast systems, there are nevertheless certain complex junctions where the latter can

245

Figure 15.23 Luminaires equipped with tubular high-pressure sodium lamps and arranged in asymmetric clusters atop 30 m masts light this motorway interchange at Wuppertal, West Germany.

Figure 15.24 Asymmetrically arranged high-mast floodlights equipped with special screening louvres light this motorway interchange at Munich, West Germany (See also colour plate 5, facing page 224).

be used to give good lighting and good visual guidance (figure 15.25). It is self evident that the decision to light a traffic junction with a certain type of high-mast lighting, with conventional lighting or even with catenary lighting (or perhaps with a combination of different systems) must be made bearing in mind the lay-out of the particular junction concerned.

Of course, economical considerations also play a role in the final decision. The cost of high-mast lighting has to be compared with that of the much larger number of masts employed with a conventional system. High masts generally call for expensive foundations, but the electricity supply is often easier and cheaper to bring in. High-mast lighting is also usually cheaper to relamp and clean than other installations due to the fewer number of lighting points involved. Finally, high-mast lighting uses light sources whose lumen output is high and whose luminous efficacy is higher than those of the lamps used in conventional installations.

Figure 15.25 A junction on the Midland Links motorway system near Birmingham, England, lighted with conventional low-mast low-pressure sodium installations: motorway, twin central, with twelve-metre masts spaced thirty-five metres apart; slip roads, single-sided, with ten-metre masts thirty-three metres apart.

Since the cost aspects are completely dependent upon the layout of the actual road network, the type of masts preferred (figure 15.26 shows a variety of different masts), local costs, local maintenance possibilities and further local circumstances, no general guidance on the best choice of installation, from the point of view of costs, can usefully be given.

a

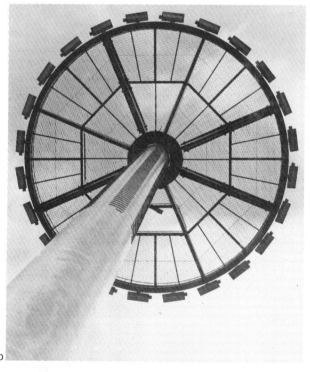

b

Figure 15.26 High-mast lighting configurations: (a) Plane symmetric, access by means of crane; (b) Radial symmetric, with access through the mast; (c) Radial symmetric, equipped with lowering gear; (d) and (e) Asymmetric, reached by climbing steps attached to mast; (f) Asymmetric, reached by crane.

c

d

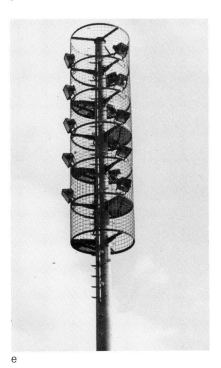

e

f

15.2 General Considerations

No part of a lighting installation, however small, should be designed in ignorance of the role it will be called upon to perform in relation to other parts of the installation adjacent to it.

Thus, while the first responsibility of the lighting engineer is to ensure that his proposed lighting scheme fully satisfies the usual quality requirements, it is his responsibility also to see to it that the question of incompatibility does not arise. The aim should be to strive for the maximum degree of harmonisation between one area and another, whether the harmony be between lighted areas or between a lighted and an unlit area.

15.2.1 Adjacent Road Lighting Schemes

Certain planned differences in design between one scheme and another can be used to improve the visual guidance given by the road lighting as a whole. One design variable frequently used toward this end is changes in the lighting system itself; another is source colour; a third is a difference in luminaire style and mounting height; and a fourth is differences in lighting configuration.

Lighting system

While it is quite common to work with just one lighting system for the various approaches to a junction and the junction itself, orientation and guidance can frequently be improved by 'picking out' the latter with the aid of a contrasting system. At complex junctions, for example, it is perhaps more common (for the reasons given earlier) to employ high-mast rather than conventional lighting, with the additional benefit that it signals to the approaching motorist that there is a junction to be negotiated.

At lesser junctions, where the main road is lighted by a catenary system, the presence of an important side road can be indicated by lighting this with a conventional luminaire arrangement. Figure 15.27 illustrates the technique employed. The visual guidance given by the catenary lighting is excellent, but at the same time the conventionally lighted exit at the right is clearly visible.

Colour

A marked difference in colour between the sources used in converging lighting schemes is a very effective way of drawing attention to a road junction, colour plate 7 facing page 225. The black and white photograph gives no clue at all to what happens to the road ahead; were the luminaires all of the same colour, all that would be seen would be a confusion of light points. But add the colour coding revealed in the colour photograph of the same situation and there, clearly bearing to the left is the main road lighted in the characteristic

Figure 15.27 This exit to the right has been clearly revealed by the conventional lighting, which contrasts well with the catenary lighting on the main road.

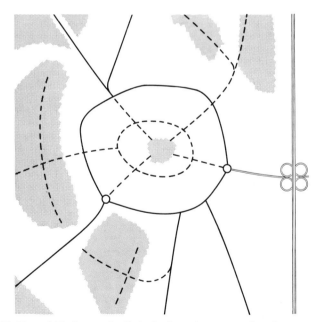

Figure 15.28 Plan of Eindhoven, the Netherlands, with route marking by means of lights of different colour (see text). Solid line denotes low-pressure sodium lighting while broken line denotes either high-pressure sodium or high-pressure mercury lighting.

251

yellowish colour given by low-pressure sodium with going off to the right an exit road picked out in white by high-pressure mercury.

The difference in colour that can be provided by road lighting can also be used at night time to supplement the information given on route indicator boards. An application of this sort is illustrated in the lighting plan of figure 15.28. Traffic wishing to avoid the middle of town is looking for a bypass or ring road along which it can turn off, and the latter has in this case been made easy to find by lighting both it and the approaches in the distinctive yellowish colour given by low-pressure sodium. All major turn-offs from the ring road leading away from the town are similarly lighted with low-pressure sodium. Traffic arriving at the ring road and wishing to continue to the town centre, on the other hand, naturally follows the white high-pressure mercury or the yellowish-white high-pressure sodium associated with town-centre approaches.

Luminaire style and mounting height
Another way of providing visual guidance at road junctions and slip-roads leading to motorway parking areas, and one that works by day as well as by night, is to create systematic differences in luminaire styling or mounting height. These techniques can also be combined as illustrated in figure 15.29.

Figure 15.29 The contrast in luminaire style and mounting height make this motorway parking area easily visible from afar.

252

Lighting configurations

Where lighting schemes converge great care must be taken to ensure that the multiplicity of lighting columns and light points visible to an approaching driver does not confuse rather than define the layout of the road junction. Figure 15.30 shows how a change in the lighting configuration, from twin-central to opposite, has been used to signal to a driver on the main road that he is approaching a dangerous crossing.

Figure 15.30 The change from twin central to opposite in the arrangement of the lighting columns on this dual carriageway near Best, the Netherlands, gives warning that one is approaching a major cross-roads.

One possible consequence of failure in this regard is illustrated by the lighting shown in figure 15.31. The road situation ahead is confused to say the least, because little or no thought seems to have been given to how the different configurations of the two neighbouring installations would be perceived by a road user looking ahead from one lighted section into the next. The line of lights gives the impression that one can drive straight on into the next section of lighting, but this is clearly impossible as the plan view of the situation shows. In short, the two lighting installations are not compatible – the one conflicts with the other.

Visual integration is obviously very important, but this must be done bearing in mind the perspective view of the road user. The perspective can be checked by way of scaled perspective drawings, one for each direction of approach to the junction. An alternative is to make use of a model that allows various lighting arrangements to be tried out in scale.

Just such a model is pictured in figure 15.32. It allows miniature 'luminaires'

Figure 15.31 Confusion through lack of planning (see text). The true situation with regard to the run of the road is revealed in the insert.

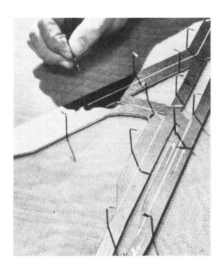

Figure 15.32 A model that facilitates the study of road lighting configurations, The miniature lighting columns are plugged in at selected positions.

to be placed in any desired configuration on a two-dimensional representation of the junction to be lighted. The luminaires (these are in fact tiny lamps) are mounted on miniature masts that plug into the model's baseboard which is covered with two sheets of metal gauze insulated from one another by a sheet of foam plastic and connected across a battery. Plugging in a mast anywhere on the board thus causes the luminaire to light. The plan of the road or road junction at the point of interest is simply a paper cut-out having a tint similar to that of the actual road surface it represents.

As the lighting arrangement has to be viewed in the right perspective, the eye height must be scaled down to correspond to the scale of the model. For the model shown here, this is done with the aid of a special viewer which can be placed at any desired point.

The lighting arrangements shown in figure 15.33 will serve to give some idea of the sort of configurations commonly adopted in practice. It will be appreciated, however, that because junctions can differ one from another on so many points it is impossible to be more specific with regard to this aspect of lighting design.

15.2.2 Terminating the Road Lighting

A road user travelling along a stretch of unlighted road and entering a section that is provided with road lighting rapidly adapts to the new conditions. A discontinuity in the lighting of this sort is therefore quite acceptable and calls for no special care in the design of the installation at the point where the road lighting commences.

Adaptation from light to dark, on the other hand, takes place comparatively slowly. A driver leaving a lighted section of road needs time to acquire the night vision that will enable him to drive safely on headlights alone. For this adaptation to take place in safety, it is necessary to terminate the road lighting gradually by providing lighting of a reduced level along the stretch of road leading into the unlighted section.

For this transition lighting to be effective it should provide a degree of perceptibility comparable with that obtained when adapted to dipped headlights alone. Research has shown (De Boer, *et al.* 1967) that an average road surface luminance of around 2 cd/m^2 as given by the normal road lighting calls for an average luminance during transition of about 0.3 cd/m^2 for this condition to be fulfilled. The transition lighting should take a driver 10 seconds or more to pass through it (e.g. for speeds of 100 km/h and 70 km/h the transition would need to take place over distances of 275 m and 200 m respectively).

The same research referred to above has also shown that there is nothing to be gained from making the transition more gradual for a given distance; as could

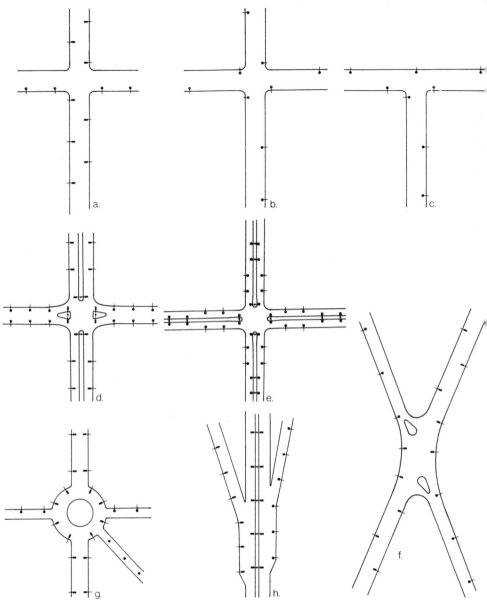

Figure 15.33 Recommended lighting configurations for a selection of commonly encountered road junctions with different symbols used to denote differences in mounting height or lamp colour: (a) Difference in column configuration accompanied by a difference in luminaire height and/or lamp type used to mark the crossing of roads of unequal importance; (b) Configuration at crossing of two major roads which are themselves lighted with either single-sided or staggered arrangements; (c) Siting of luminaires at T-junction highlights danger area (note especially luminaire opposite the base of the T); (d) Change in lighting configuration on dual carriageway and difference in luminaire height or lamp type warns of crossing; (e) Symmetrical lighting configuration at crossing of two equally important dual carriageways, with change from twin-central to opposite to warn of approach; (f) Crossing of major and minor roads revealed by difference in mounting height or lamp type; (g) Roundabout lighted using luminaire height/lamp combination employed on main approaches; (h) Motorway slip-roads, with contrasting luminaire height or lamp type used to mark exit.

8 *These 150 watt incandescent lighting bollards have been placed to clearly reveal the steps leading to this sunken garden, without dispelling the night-time seclusion of the surroundings. (See Sec. 16.2.3)*

9 A pleasantly lighted pedestrian area where the comfort of the neighbouring residents in their homes has not been o
looked. Unwanted stray light in the direction of the houses has been kept to an acceptable level by employing downligh
having a very pronounced cut-off above the horizontal. (See Sec. 16.2.2)

10 The Gräfelfing tunnel near Munich, West Germany. Seen here are some of the triple-lamp low-pressure sodium a
twin-lamp fluorescent luminaires providing the high level daytime lighting in the threshold zone. As outside levels fall i
lighting is automatically reduced in six steps until, at night, only a single row of fluorescents is left burning. (See Sec. 17.1

be done, for example, by providing a number of luminance steps. It is best, therefore, to keep the column spacing and luminaire mounting height constant and reduce the lighting level by employing lamps of a lower power rating. An alternative solution (Boereboom, 1966) where an opposite or twin-central arrangement of the columns is to be employed along the normally lighted section, is to simply light the traffic lane leading into the unlighted zone from the side of the road farthest from it (figure 15.34).

Figure 15.34 Change from twin central to back-lighting as a means of terminating the lighting on a dual carriageway.

15.2.3 Lighting and the Environment

There is one final side to road lighting that needs to be considered before discussion on the general aspects of design can be brought to a close. This concerns the importance of ensuring that a lighting scheme is not only acceptable to road users, but to other members of the public and travellers employing other means of transportation as well. (The lighting of residential and pedestrian areas is dealt with in the following chapter.)

Any serious glare disturbance to non road users must, of course, be avoided. Luminaires having tightly controlled light distributions can be employed to minimise glare at critical locations; for example, where a lighted road bridge passes over a road, or where luminaires are perforce sited close to houses or other premises that are occupied after dark.

For tighter control, special screens or louvres may provide the answer. In the case of the lighted bridge shown in figure 15.35, for example, screens (figure 15.36) have been fitted to the luminaires in order to avoid destroying the night vision of ships' pilots on the river below. Parapet lighting (figure

Figure 15.35 The Bosporus bridge, Istambul, Turkey, The bridge supports are floodlighted, while the road itself is lighted using well screened high-pressure sodium luminaires (see figure 15.36).

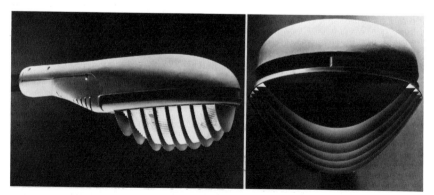

Figure 15.36 The luminaire employed in the Bosporus bridge installation, showing the special louvres used to screen the lamp from view from the river below.

258

Figure 15.37 A road bridge near São Paulo, Brazil, provided with parapet lighting because it was felt that conventional road lighting columns would have interfered with the view of the adjacent traffic junction.

15.37) would seem to offer an attractive alternative where bridges over navigable rivers are concerned as such lighting is automatically screened from below by the parapet. However, the required lighting level and uniformity are difficult to achieve and maintain with this form of lighting; it is therefore not to be recommended, except where conventional mounting heights are impracticable, or undesirable from an aesthetic point of view.

Some sort of luminaire screening can also be employed where it is necessary to 'black out' certain luminaires in order to prevent their creating a confusing pattern of light points when viewed from some way off. Unshielded street lights, for example, are easily confused with those marking a harbour entrance when seen from out at sea. Similarly, when viewed from the air such lights may conflict with the runway lights of a nearby aerodrome.

Chapter 16

Residential and Pedestrian Areas

The road lighting covered so far has been centred on the needs of motorised traffic. In those parts of residential areas where motorised traffic is allowed, the lighting should be aimed at promoting road safety on the one hand and security and amenity on the other. In pedestrian areas closed to all road traffic, road safety ceases to be a design consideration and full scope can be given to providing effective security and amenity lighting.

The lighting in both residential and pedestrian areas should, and can if properly designed, increase their attractiveness to local residents and visitors alike. This challenge can be met by architects, town planners and lighting engineers working together, each drawing on the knowledge and experience of the other.

In this chapter the lighting needs of pedestrians and local residents are reviewed and the lighting criteria necessary to their fulfillment discussed. Finally, the lamps, luminaires and lighting arrangements available to put theory into practice are outlined.

16.1 Lighting Criteria

The demands made on the lighting by pedestrians and local residents can be summarised as follows

pedestrians: that it should facilitate movement and orientation and promote facial recognition
local residents: (in their homes): that it should help to reveal the presence of intruders and not constitute a disturbance (especially in the form of glare)
both groups: that it should enhance the attractiveness of the surroundings and be sufficiently functional to discourage violence, vandalism and crime.

The lighting criteria of primary importance in the fulfillment of these demands are: lighting level (illuminance), uniformity (specifically illuminance uniformity) and glare restriction.

16.1.1 Lighting Level

The recommendations for the lighting of public footpaths and pedestrian areas can be compared with those for the lighting of roads in so far as they are both designed to facilitate movement and recognition at night. There are two main points of difference. The first is that compared with motor vehicles, pedestrians move relatively slowly and this means that the human eye has more time to adapt to changes in brightness. Lighting levels and uniformity, particularly the latter, are therefore less critical than is the case with faster moving road traffic. The second difference is that whilst the driver of a motor vehicle is not wholly reliant on road lighting for his orientation – he has his vehicle headlights to help him – a pedestrian has only what illumination is provided along the way; the minimum value of such lighting is therefore of great importance.

Safe movement. It is important for pedestrians to be able to move about safely, so lighting levels should be sufficient to reveal potentially dangerous obstacles lying in their path and any irregularities in this. These requirements will be met with a horizontal illuminance of 1 lux or greater, this being the value recommended in the CIE Guide on the Emergency Lighting of Premises (CIE, 1979).

Since most objects are not flat but three dimensional, it is sometimes preferred to specify the strength of the lighting in terms of the hemispherical illuminance rather than the horizontal illuminance. In Denmark (DEN, 1979), the values recommended for the former for various categories of residential road range from 1 to 5 lux. Cylindrical illuminances have also been proposed, for the same reason (Epaneshnikov, *et al*., 1971). (The terms vertical, cylindrical and hemispherical illuminance are defined in figure 16.1.)

E_v E_c E_{hs}

Figure 16.1 The terms vertical illuminance (E_v), cylindrical illuminance (E_c) and hemispherical illuminance (E_{hs}) refer to the illuminance falling on a small body of the shape illustrated.

Facial recognition

It is important for pedestrians to be able to recognise one another on meeting. A recent study (Caminada and Van Bommel, 1980) shows that the lighting parameter best correlated with facial recognition is the semi cylindrical illuminance. From tests conducted outdoors under lighting typical for a residential area it was found that for facial recognition at an observer distance of 4 m (the 'safe' distance if attack is threatened – see *Security*, below) a semi cylindrical illuminance of 0.8 lux on the face was needed.

Orientation. Good orientation implies the ability to identify houses and other buildings and features of the environment, especially road junctions. These requirements will be met as a matter of course with most types of residential lighting. Street names in particular should be well illuminated.

Security. Residential lighting normally has a twofold security function to perform: it should help to deter a would-be intruder or burglar, but failing in this it should at least make his presence known to residents and passers by. Both these aims will usually be met if the requirements concerning facial recognition are fulfilled (Caminada, 1979).

Good lighting will serve not only to reduce the number of night-time break-ins, it will also prove a strong deterrent to street crime and vandalism (Fischer, 1978b).

The pedestrian is especially vulnerable to attack after dark. To combat this it is to be expected that the lighting should at least facilitate facial recognition from a distance at which avoiding action on the part of the person threatened is still possible. Tests have shown (Caminada and Van Bommel, 1980) that for safety in this respect, the requirement relating to facial recognition specified above (viz. a semi cylindrical illuminance on the face of about 0.8 lux) is again applicable.

Recommended levels

From relations existing between horizontal, hemispherical (both at ground level) and vertical illuminance (at a height of 1.5 m), Fischer (1978c) summarised the various national and international recommendations in the form of a table of recommended horizontal illuminances for residential areas (table 16.1) – road safety was not a relevant consideration. From what was said above concerning facial recognition, however, it would perhaps be better to

Table 16.1 Recommended horizontal illuminances for residential areas (Fischer)

Illuminance	Remarks
1 lux (minimum)	Minimum in order to positively detect obstacles
5 lux (average)	Facilitates positive orientation
20 lux (average)	Attractive lighting
	Human features are recognisable

base future recommendations on a minimum value for the semi cylindrical illuminance.

Where road safety is a consideration, the recommendations given in Chapter 9 become applicable.

16.1.2 Illuminance Uniformity

Problems of visual adaptation among pedestrians will be avoided if the ratio of the maximum to the minimum illuminance does not exceed about 20:1 (Fischer, 1978c). Again, if road safety has to be considered, the luminance uniformity recommendations given earlier in this book should be followed.

The degree of uniformity, or rather non-uniformity, specified above gives the designer the freedom necessary for the creation of an imaginative lighting scheme (colour plates 6 and 8 facing pages 224 and 256); lighting that is too uniform will inevitably produce a dull overall effect and so should be avoided.

16.1.3 Glare Control

The problem of glare is less critical for pedestrians than it is for motorists. This is primarily because of the much lower speeds involved where pedestrians are concerned. The pedestrian has far more time in which to adapt to changes in brightness in his visual field and is therefore less likely to be so

Figure 16.2 Some brilliance from the luminaires providing the outdoor lighting helps to create an attractive and animating effect in this pedestrian shopping area.

263

blinded as to come into collision with an unseen obstacle in his path. In fact some brilliance is often welcomed as it can help to create an attractive and animating effect (figure 16.2).

An important rule in keeping glare to an acceptable minimum is that no unscreened light sources should be located at eye level; they should either be lower than about one metre from the ground – as is the case with lighting bollards – or higher than about three metres.

A pedestrian is more likely to be troubled by discomfort glare than by disability glare. The validity of the glare control mark (G) – the measure for discomfort glare – has not been investigated for luminaires mounted lower than 6.5 metres above ground level. Furthermore, even where the mounting height is greater than this value, the glare control mark can seldom be used in recommendations involving residential areas because here the luminaire layout is invariably irregular and G can only be determined for regular rows of luminaires.

In practice, however, the glare sensation will more probably be caused by individual bright luminaires appearing near to the direct line of sight than by the combined effect of all the luminaires in the area within view. It is sensible, therefore, to limit the luminance of the individual luminaires for critical angles of emission.

In Denmark (DEN, 1979) the luminaire parameter $I/\sqrt{A}$ is used for this purpose.

Here I is the luminaire's luminous intensity in candela and A its flashed (or light emitting) area in square metres in the direction $85°$ from the downward vertical. The ratio $I/\sqrt{A}$ may be written as the product $L\sqrt{A}$, where L is the luminance of the luminaire when viewed from the same direction. The fact that the value of this product is specified and not the luminaire luminance alone means that the larger the light emitting surface of the luminaire (i.e. the larger the area occupied by this in the pedestrian's field of view), the lower must be its luminance.

Caminada and Van Bommel (1980) investigated the influence of this light emitting area and its luminance on the glare appraisal returned by a large group of observers in a normally lighted residential environment. This study shows that the product $LA^{0.25}$ is the luminaire parameter best suited for the control of glare in residential areas, and that the permitted maximum for this parameter is governed by the luminaire mounting height, h, as follows

for $h < 4.5$ m $\quad\quad LA^{0.25} < 1250$
for 4.5 m $< h < 6$ m $LA^{0.25} < 1500$
for $h > 6$ m $\quad\quad LA^{0.25} < 2000$

where L is in cd/m^2 and A is in m^2.

The permitted luminances for some typical luminaires employed in residential areas, as calculated from these requirements, are given in table 16.2.

264

Table 16.2 *Permitted luminances of some typical luminaires for a viewing angle of 85° from the downward vertical.*

Type of luminaire	Area at 85° (m²)	Permitted luminance (cd/m²)		
		$LA^{0.25} \lessgtr 1250$	$LA^{0.25} \lessgtr 1500$	$LA^{0.25} \lessgtr 2000$
General diffuse sphere: $d = 0.50$ m	0.20	1870	2240	2990
General diffuse sphere: $d = 0.60$ m	0.28	1720	2060	2750
Conventional road lighting, reflector	0.025	3140	3770	5030
Conventional road lighting, refractor	0.035	2890	3470	4620

16.2 Lighting Installations

16.2.1 Lamps

The lighting installations in residential areas commonly employ the following lamp types: incandescent, blended-light, tubular fluorescent, high-pressure mercury, high-pressure sodium and low-pressure sodium. (High-pressure mercury and high and low-pressure sodium lamps are also employed in road lighting, and their characteristics are dealt with in some detail elsewhere in this book – see especially Chapters 6 and 12.)

Incandescent. The incandescent lamp with its pleasant, warm-coloured light is still sometimes used in residential areas, but its comparatively high energy consumption and short life are prompting an ever increasing number of authorities to go over to discharge lamps.

Blended-light. The blended-light lamp – a combined incandescent and high-pressure mercury lamp needing no external ballast – has a longer life and a higher efficacy than incandescent lamps and since it needs no control gear it is a lamp often used to replace incandescent lamps in older installations.

High-pressure mercury. The high-pressure mercury lamp is slightly cooler in colour than the blended-light lamp, but has a far higher efficacy. It is available in a wide range of lumen outputs, including the lower outputs usually needed for residential areas. It is currently the most frequently employed lamp type in these areas.

Tubular fluorescent. Of the great variety of tubular fluorescent lamps available, only the 20 watt linear lamp and those of the circular or *U*-shaped

265

varieties in the range 20 to 40 watts are compact enough to be used in the relatively small luminaires desirable for residential area lighting. The tubular fluorescent lamp is therefore often used where low lighting levels are required, which may be the case in certain parts of a residential area.

High-pressure sodium. The efficacy of the high-pressure sodium lamp is ten times that of the incandescent lamp and almost double that of the high-pressure mercury lamp. This lamp, which gives an even warmer light than the incandescent lamp, is therefore a very suitable lamp for use in residential areas, especially since it has recently become available in low-wattage, low output versions.

Low-pressure sodium. The low-pressure sodium lamp has the highest efficacy of all lamp types, but because of its absence of colour rendering is not favoured in residential or pedestrian areas.

16.2.2 Luminaires

There is a great variety of luminaires now available for the lighting of residential and pedestrian areas differing in shape, in light distribution, and in styling.

Luminaire style and shape are major factors in determining the day (and often the night-time) appearance of the lighting installation. It is therefore very important that this be kept in mind when striving for a certain overall impression.

But it is the light distribution, shape and dimensions of the luminaire which are really important at night. These factors determine the brightness of the luminaire and the distance apart at which these can be spaced in a given luminaire arrangement.

Where the lighting installation must also serve the needs of the motorist, conventional road lighting luminaires are usually the best choice from the economical point of view. Even such functional luminaires can be made to harmonise with their surroundings (figure 16.3).

There is of course more freedom in the choice of the luminaire type when there is only little or no motorised traffic. Some idea of the extremes of light distribution that can then be used is given by way of figure 16.4, which shows two typical rotationally symmetric luminaires together with those characteristics that determine their possible areas of useful application. The examples are taken from the German 'Guide for the Lighting of Pedestrian Areas' (FfdS, 1977).

It can be seen that the one extreme of light distribution, as given by the general-diffuse type of luminaire, is suitable for the lighting of relatively large

Figure 16.3 Conventional road lighting luminaires serve the needs of motorist and pedestrian in what is essentially a residential area.

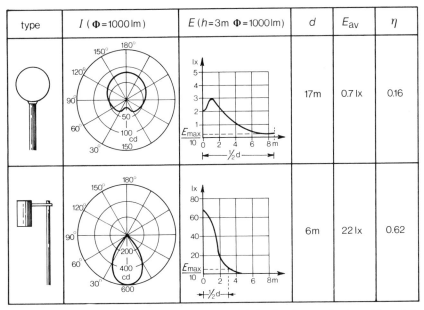

type	I (Φ=1000 lm)	E (h=3m Φ=1000 lm)	d	E_{av}	η
			17m	0.7 lx	0.16
			6m	22 lx	0.62

Figure 16.4 Luminaires exhibiting two extremes of light distribution: (top) a globe of the general-diffuse type, and (bottom) a downlighter. Both luminaires are mounted three metres above the ground and each has an output of 1000 lumens. Also given are: d, the diameter of the circle at ground level within which $E_{max}/E_{min} = 10$; E_{av}, the average illuminance within this circle; and η, the utilisation factor for the circle.

267

areas to a rather low level of illuminance. The vertical and semi cylindrical illuminances will be relatively high with luminaires of this type, a fact that may be especially useful to remember where facades adjacent to the lighted area must also be illuminated.

The luminaires giving the other extreme of light distribution, the direct luminaires, or 'downlighters', have to be spaced at rather short intervals if adequate uniformity is to be obtained. The horizontal illuminance directly below the luminaire will be relatively high and the danger of glare reduced considerably. On the other hand, this means that vertical and semi cylindrical illuminances will be low (colour plate 9 facing page 257).

16.2.3 Luminaire Mounting

The lighting in residential areas is usually provided by luminaires mounted in one or more of the following four positions: atop posts or columns, on facades, suspended overhead, and on the ground.

Post and column-mounted luminaires

The post-top luminaire, by which is usually meant a luminaire mounted atop a post (no support arms) some three to eight metres tall, is by far the most widely employed luminaire mounting in residential areas.

The mounting height employed for post-tops is largely dependent upon the area of the surface to be illuminated. Post-tops modelled in the modern decorative style or in the style of the old gas lanterns are normally called upon to illuminate only the pathway in their immediate surround and are therefore mounted no more than three to five metres above the ground. Where the area to be lighted is extensive it is more usual to find that the luminaires used are functional rather than decorative and that they are mounted somewhat higher, say between five and eight metres from the ground.

Facade-mounted luminaires

Lighting given by facade-mounted luminaires provides a very good and attractive solution to the problem of how to light streets where there is no space for columns (colour plate 6 facing page 224), or areas bordered by historic buildings that should not be obscured by lighting masts.

This ancient form of lighting can also be used to great effect in modern civic centres. Decorative luminaires mounted so that their shape is clearly revealed can add to the charm of the surroundings (figure 16.5). Alternatively, the luminaires can be hidden from view, a few well directed optically equipped units being used to unobtrusively wash an attractive facade with light. This wall-wash technique, which in the effect created is similar to that obtained using floodlighting, offers many possibilities for artistic expression.

268

Figure 16.5 Modern luminaires on the wall of the New York State Theatre in the Lincoln Center, New York.

Suspended, overhead luminaires

Where an area must be kept free of lighting columns and where facade mounting is undesirable (perhaps because it would illuminate this too strongly) the luminaires can be suspended overhead from wires anchored to the adjacent buildings. Luminaires suspended overhead in this way can be used to create a sort of lighted roof to the scene below and can endow it with a rather special, intimate character.

Ground-level luminaires

In parks and gardens and along public footpaths, where the only purpose is to provide atmosphere and orientation, it is often possible and advantageous from an aesthetic point of view to mount the luminaires close to the ground in special lighting bollards (colour plate 8 facing page 256).

Chapter 17

Tunnels and Underpasses

The fundamentals of tunnel lighting are discussed in detail in Part 1 (Chapter 5) of this book. The present chapter goes a step further and considers how this knowledge can best be applied in practice.

The fact that there will generally be more than one lighting design that is acceptable for a given situation, means that ultimately a choice will have to be made. This choice will often depend upon constructional design details of the tunnel itself, although cost and energy considerations also inevitably play an important role. In this chapter, various practical tunnel lighting schemes together with details of individual designs will be discussed.

Much can be done by way of relatively simple measures planned during the tunnel design phase to keep lighting costs to a minimum. These measures and the important question of lighting maintenance are also considered.

17.1 Day-time Lighting

17.1.1 Threshold Zone Lighting

To make obstacles just inside the tunnel visible to the approaching driver it is necessary to increase the luminance level of the tunnel entrance or threshold zone, for the amount of daylight penetrating the tunnel entrance is so small that it can be neglected (figure 17.1).

Furthermore, what daylight does reach the road surface is coming from above and behind the approaching driver, so the increase in road surface luminance attributable to it will be very small indeed.

The threshold zone lighting must therefore come either from artificial lighting inside the tunnel or, alternatively, an artificial threshold zone can be created in front of the tunnel entrance proper, which can then be lighted using daylight screens. Whichever approach is adopted, and examples of each will be considered below, the actual luminance needed in the threshold zone will be dependent upon the outside (equivalent) adaptation luminance prevailing. (The relation between outside adaptation luminance and the luminance needed in the threshold zone is discussed in detail in Chapter 5, as is the length

270

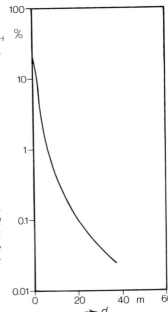

Figure 17.1 Daylight penetration at a tunnel entrance. The illuminance on the road inside the tunnel (E_H) as a percentage of that on the road just in front of the entrance is given as a function of the distance in from the tunnel entrance. Dimensions of entrance: 4.5 m high by 10 m wide. (Schreuder)

needed for this zone. This discussion is summarised in Sec. 9.2, where the recommendations applicable to tunnel lighting are outlined.)

Outside adaptation luminance

In the case of existing tunnels, where the luminances in the field of view can be measured, the adaptation luminance – or rather the equivalent adaptation luminance – can be approximated (See Chapter 5). With new tunnels, on the other hand, the adaptation luminance must be known some time before work on the tunnel is completed so that the lighting design can be finalised and the scheme installed prior to the tunnel being opened.

For this purpose, investigations aimed at producing a detailed classification of tunnel types according to the maximum adaptation luminance likely to be found at the entrance are in progress (CIE, 1979b).

Naturally, the adaptation luminance does not remain constant; it is continually changing in response to variations in the amount of daylight present. The maximum adaptation luminance is likely to correspond to a horizontal illuminance of 100000 lux, this being about the highest value occurring outside under normal conditions. (Higher maximums are, of course, always possible: particularly high adaptation luminances can occur, for example, at sunrise or sunset where a tunnel in the open country is east-west oriented, or indeed at any time of the day for a mountain tunnel the surrounds of which are covered by snow.)

271

Figure 17.2 The dark-coloured walls of this approach cutting, the dark façade of the tunnel entrance, and the newly planted trees seen against the skyline all serve to lower the effective outside adaptation luminance and so permit of lower lighting levels in the tunnel entrance zone.

Figure 17.3 This large entrance canopy, by shielding a large section of sky from view, effectively limits the outside adaptation level and so allows the lighting levels in the tunnel entrance zone to be that much lower.

Special measures can be taken and may be in force to limit the maximum adaptation luminance that can occur. Such measures include: employing non-glossy dark-coloured materials for the road surface at the tunnel approach, the tunnel entrance façade and (where applicable) the walls of the approach cutting (figure 17.2); planting trees or shrubs adjacent to and (more important) above the tunnel entrance (see again figure 17.2); and making the tunnel entrance itself as high and as wide as possible (figure 17.3).

272

Artificial lighting

Where a threshold zone is provided with artificial lighting, there are three main design variables to consider: the luminous intensity distribution of the luminaire(s); the number and geometric arrangement of these; and the type of light source(s) employed.

Any luminaire employed in tunnel lighting can in fact be broadly placed in one of two groups according to whether its luminous intensity distribution in a vertical plane parallel to the tunnel axis is symmetrical or asymmetrical (figure 17.4). Lighting installations employing luminaires of the former variety are usually referred to as being 'conventional', the term 'counter cast' being aptly used to describe installations in which the luminous intensity has its maximum in the direction towards oncoming traffic.

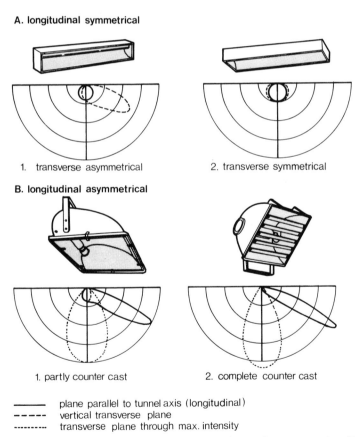

A. longitudinal symmetrical

1. transverse asymmetrical   2. transverse symmetrical

B. longitudinal asymmetrical

1. partly counter cast   2. complete counter cast

——— plane parallel to tunnel axis (longitudinal)
- - - - vertical transverse plane
.... transverse plane through max. intensity

Figure 17.4 The luminous intensity distributions corresponding to the two categories of luminaire employed in tunnel lighting.

Conventional lighting. A conventional installation may consist of several rows of luminaires concentrated above the kerb of the carriageway (often wall mounted, figure 17.5), each luminaire having a strongly asymmetric luminous intensity distribution across the tunnel (figure 17.4A.1). Alternatively, the light may come from rows of luminaires concentrated above the centre of the carriageway (figures 17.6), the luminous intensity distribution across the road in this case being symmetric (figure 17.4A.2), or perhaps slightly asymmetric.

Figure 17.5 A tunnel threshold zone lighted by wall-mounted luminaires of the conventional, transverse asymmetrical type.

The efficiency of the lighting installation is largely dependent upon which of these luminaire arrangements is employed. Calculations embracing both types of lighting arrangement and different luminous intensity distributions (Westermann, 1975a) indicate that the central arrangement can be some 25 to 50 per cent more efficient in terms of road-surface luminance yield (cd/m^2 per lumen installed per metre length of tunnel) than the side arrangement, in which the luminaires are wall mounted. Actual measurements made on existing tunnel installations, in which the same yardstick of efficiency was employed, show the conventional central arrangement to be between 40 and 70 per cent more efficient than the conventional side arrangement in which the luminaires are ceiling mounted (Walthert, 1977b) and on average 30 per cent more efficient than side arrangements in general (Müller and Riemenschneider, 1975).

274

Figure 17.6 Conventional, transverse symmetrical luminaires above the centre of the carriage-way in the threshold zone.

The obvious first choice is thus the central arrangement. It may be, however, that the height of the ceiling is insufficient to permit of such an arrangement – a minimum clearance height above the road is always specified – or that the ceiling void is needed for ventilation purposes. In such cases the ceiling-mounted side arrangement is probably the next best choice, for with the wall-mounted arrangement it is difficult to provide sufficient luminance low down on the walls themselves, and adequate wall luminance is important (see Chapter 5).

Another factor influencing the efficiency of the installation is, of course, the type of light source employed. The low-pressure sodium lamp, the source with the highest luminous efficacy, has a shape that lends itself well for use in the luminaires employed to provide conventional tunnel lighting (figure 17.6 and colour plate 10 facing page 257).

It is therefore invariably the most efficient to use source for threshold zone lighting. The high-pressure sodium lamp, however, can sometimes be employed to advantage in situations where the required lighting level is relatively high (as is sometimes the case with tunnels under rivers). Its luminous flux is higher than that of the low-pressure sodium lamp and the luminaire housing it is smaller, and this means that fewer luminaires are needed and that less space is occupied by the lighting (figure 17.7).

It is not always possible for the optics of a luminaire equipped with a high-powered lamp to prevent direct light from this being emitted in the direction

Figure 17.7 Threshold zone lighting employing space-saving high-pressure sodium luminaires, with a continuous row of tubular fluorescent luminaires running the full length of the tunnel.

of approaching traffic. Intolerable glare can be avoided in such a case by fitting the luminaire with small internal louvres (figure 17.8).

The use of tubular fluorescent lamps in the threshold zone will result in the lighting having a very low efficiency. This is because of the low efficacy and low luminous flux of the fluorescent lamp compared with low and high-pressure sodium lamps, a large number of lamps and bulky luminaires being needed to provide adequate lighting (figure 17.9). However, because such luminaires are ideally suited for use in the interior zone of the tunnel where lighting levels are somewhat lower and fewer luminaires are needed, it is generally considered advisable to continue one row of these through the transition zone and into the threshold zone for visual guidance purposes (see again figures 17.6 and 17.7). These same luminaires can then be used to provide homogeneous night-time lighting throughout the tunnel (the special requirements of which are discussed in the second part of this chapter).

Counter-cast lighting. The idea of using counter-cast luminaires for lighting tunnels is not new (Mäder, 1969). By directing the maximum amount of light towards oncoming traffic they ensure a high value for the road-surface luminance coefficient (viz. large L/E value). However, the fact that this does not automatically result in an improvement in luminance yield compared with conventional installations is due to the necessity of having to take special measures to prevent glare, the efficiency of the luminaires being sacrificed in the process.

276

Figure 17.8 Close-up of the high-pressure sodium luminaires shown in figure 17.7 The built-in louvres are clearly visible.

Figure 17.9 Where the comparatively high level threshold zone lighting is provided by tubular fluorescent lamps, these must necessarily be employed in very large numbers.

Figure 17.10 Glare from partly counter cast luminaires is avoided by mounting these with the front glass close to the horizontal.

Figure 17.11 The fully counter cast luminaire employed in this installation has a built-in louvre to obviate glare.

Counter-cast lighting has both good and bad points so far as its effect on visibility is concerned. The fact that little or none of this directional lighting reaches objects on the road from the side of an approaching driver, means that these stand out in silhouette against the bright road surface or wall beyond. Contrast between background and object is high, and this makes the latter clearly visible.

By the same reasoning, however, objects not thrown into silhouette in this way are virtually impossible to detect.

This will be the case, for example, if an object is hidden in the shadow cast by a large van, or if it merges with the unlit rear of the van itself. This situation will obviously be worse for fully counter-cast lighting (figure 17.4B.1) than for the partly counter-cast system (figure 17.4B.2): the latter gives some light in the direction of travel and this helps to keep object backgrounds at least partly illuminated.

Both the partly counter-cast and fully counter-cast luminaires are designed to be mounted on the central part of a tunnel's ceiling. Glare from the partly counter-cast version is usually prevented by mounting it with its front glass close to the horizontal (figure 17.10). The fully counter-cast unit (figure 17.11) has special built-in louvres which serve the same purpose (figure 17.4B.2).

The most suitable light source for both versions of this luminaire is the tubular high-pressure sodium lamp, the compact form of which facilitates the design of the narrow-beam optical systems involved.

Some idea of the comparative efficiency of counter-cast and conventional installations can be gained by interpreting the results of measurements made on completed tunnel lighting installations (Walthert, 1975b and Müller/Riemenschneider, 1975). The measurement results as originally presented, showing the percentage improvement in efficiency of counter-cast lighting over central-ceiling conventional lighting based on the use of the high-pressure sodium lamp throughout, are given in the first half of table

Table 17.1 Overall luminance yield of counter cast tunnel lighting installations as a percentage of that given by conventional ceiling mounted installations

	Same lamp type (from original publications)		Transformed for most efficient lamp type*	
	part counter cast	complete counter cast	part counter cast	complete counter cast
Walthert	+ 65 %	− 5 %	+ 30 %	− 25 %
Müller/Riemenschneider	+ 40 %		+ 10 %	

* for counter cast: tubular high-pressure sodium
  for conventional: low-pressure sodium

17.1. The improvement in lighting efficiency, assessed in this way, is impressive: as much as 65 per cent in the case of the partly counter-cast system. But these results could be misleading, for they apply to conventional lighting as given by the high-pressure sodium lamp and not the low-pressure sodium lamp which is more efficient. A more realistic comparison is obtained when the most appropriate lamp type is employed for each system, viz. HPS for counter-cast lighting and LPS for the conventional. The results of such a comparison are given in the second half of table 17.1. As can be seen, the improvement in efficiency given by counter-cast lighting is now low to moderate, and in some cases where fully counter-cast lighting is concerned there may even be a drop in efficiency compared with conventional lighting.

Switching

The luminance given by the lighting in the threshold zone of a tunnel must at all times be an appropriate specified percentage of the outside adaptation luminance. This calls for the provision of automatic switching facilities that allow the level of the artificial lighting to be adjusted according to requirements.

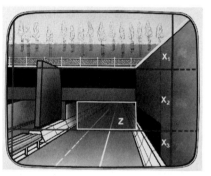

Figure 17.12 Automatic control of lighting level. A closed-circuit television camera directed toward the tunnel entrance monitors the changing brightness in its field of view. The camera generates a so-called luminance signal, the instantaneous amplitude of which is proportional to the luminance of the picture element being scanned. The weighted average of the integrated luminance signal corresponding to picture areas X_1, X_2 and X_3 is electronically compared with the similarly processed signal corresponding to area Z. According to the degree of balance obtained from this comparison, so a section of the tunnel lighting is switched on or off.

The switching may be done in a stepwise manner, the reduction to lower levels and the increase to higher levels being made in steps of no greater than 3 : 1 and 1 : 3 respectively (CIE, 1973b).

It is advisable to arrange for at least one continuous row of luminaires to be left burning to provide the minimum luminance needed. This ensures that good visual guidance is maintained and that there will be no flicker.

If only sodium lamps are used in the threshold zone, the lower lighting levels have perforce to be obtained by switching off alternate luminaires along a row. Special care should then be taken with regard to the spacing of the luminaires in order to avoid producing disturbing flicker.

The switching in many older tunnels is performed in response to signals received from photocells, which continuously monitor the illuminance level at each entrance. It is preferable, however, to monitor the outside adaptation luminance direct, or at least some quantity closely related to it, which illuminance is not. In some recently completed tunnels, therefore, use has been made of new techniques in which specially adapted television cameras (Van den Bijllaardt, 1977) and film cameras (Schröter, 1977) have been employed toward this end. Each camera (figure 17.12a), one at each entrance, is mounted so as to obtain the approximate field of view of an approaching motorist (figure 17.12b), so the luminance measured and used for control purposes is very closely related to the adaptation luminance actually prevailing at a given instant.

Daylight screens

As was mentioned earlier, an alternative to increasing the luminance within the tunnel entrance is to create an exterior threshold zone by the use of daylight screens. The screens, erected immediately outside the entrances to the tunnel, control the amount of daylight reaching the road according to the intensity of the daylight incident upon them.

Three fundamentally different types of screen have been tried out in practice: semi-open, allowing no direct sunlight to penetrate; open, allowing direct sunlight to penetrate; and closed, working by diffuse transmission only.

A semi-open screen is shown in figure 17.13. These 'suntight' screens were built originally in the form of heavy concrete louvres, but nowadays light-weight aluminium structures are more common. A critical point in their design is the size and shape of the openings: these must be small enough to prevent direct sunlight from penetrating for the height involved, but not so small that they can become clogged by snow as this would shut off too much light. The louvre size necessary to satisfy the first of these conditions can be determined graphically (Schreuder, 1964).

Since part of the light transmission of these screens is obtained by means of reflection from the vertical surfaces, their overall transmission will be considerably influenced by soiling of these surfaces, which are very difficult to clean. Another disadvantage of these screens (Van den Bijllaardt, 1977) is that the transmission also varies with the varying sky condition (figure 17.14).

In regions subject to sub-zero temperatures, rain or snow passing through the screen and falling on the road surface below can easily freeze without the warming effect of direct sunlight. Where this is a possibility, some form of road heating is recommended.

Daylight screens (whether these be horizontal and situated above the approach road or vertical and placed next to it) that allow direct sunlight to reach the road surface, are not recommended for three reasons. In the first place, the very fact that they allow sunlight to strike the road direct, and thus possibly a driver's eyes as well, means that there is every chance that a driver

Figure 17.13 A daylight screen of the semi-open variety, which prevents direct sunlight from reaching the road surface and a driver's eyes.

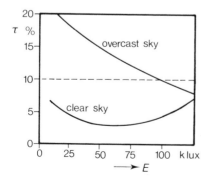

Figure 17.14 Variation in transmission with change in horizontal illuminance for a daylight screen (consisting of aluminium louvres) of the semi-open variety. The illuminance on the screen, at the Velsen tunnel (the Netherlands), was measured with change in the sun's elevation (Van den Bijllaardt)

will be seriously troubled by flicker. Secondly, there are liable to be disturbing patterns of light and shade on the road surface. Thirdly, their transmission varies enormously in response to both the condition of the sky and the altitude of the sun in it.

This leaves the daylight screens of the totally closed variety, which are constructed of semi-translucent plastics or glass. Again, these are not to be recommended. They are perforce expensive in construction, and transmission is very strongly influenced by soiling because these screens, unlike the others considered, rely 100 per cent on diffuse transmission for their proper working.

17.1.2 Transition-zone Lighting

A driver entering a long tunnel needs time for his eyes to become sufficiently adapted to the low-luminance central zone. The transition from the highest to the lowest luminance levels present must therefore be made gradually, and this is the purpose of the lighting in the transition zone (see Chapter 5 and Sec. 9.2).

In practice, the luminance is not reduced smoothly along the transition zone (as is suggested by the smooth luminance reduction curve given in Chapter 5) but in a number of steps. The luminance ratio between one step and the next must not exceed $3:1$ and, depending on the value of the threshold-zone luminance, this will usually mean having between 2 and 4 steps to complete the transition.

The lighting scheme for the various sections of the transition zone can, in principle, be the same as that provided in the threshold zones. The necessary reduction in lighting level at each section can be brought about by employing lamps of lower lumen output, by reducing the number of rows of luminaires, or by a combination of both.

17.1.3 Interior-zone Lighting

The lighting levels needed in the interior zone of a tunnel are detailed in Sec. 9.2. The luminaires providing this lighting should preferably be placed in a continuous row so as to avoid flicker. Discontinuities in the rows of luminaires are, however, permissible provided the necessary steps are taken to ensure that the flicker so created is not disturbing (see Sec. 5.5).

The most suitable luminaire type for use in the interior zone is that housing the tubular fluorescent lamp. A row of these should be continued through into the transition and threshold zones, as this helps to provide visual guidance; this may mean that the lamps will have to be dimmed in order to give the relatively low night-time level.

17.1.4 Exit Lighting

In the daytime, the exit of a tunnel appears to a driver inside as a 'bright hole' against which obstacles can be clearly seen in silhouette. The silhouette effect is created in part by virtue of the daylight being incident on the walls, ceiling and road against the direction of view so creating high luminance coefficients at these surfaces. The effect can be enhanced and extended farther into the tunnel to give a more gradual increase in brightening by making use of surface materials having a high degree of reflectance.

As visual adaptation from a low to a high level of luminance takes place rapidly, the requirements for exit lighting are far less severe than those for the entrance. There is, however, an advantage to be gained in making the entrance and exit lighting symmetrical: it is then possible, in the event of a traffic blockage or during periods of maintenance, to use a single tunnel-tube to carry two-way traffic.

17.2 Night-time Lighting

At night-time, the supplementary lighting used in the various zones to meet day-time requirements has to be switched off and the remaining lights either reduced in number or dimmed. With regard to the entrance and exit, the day-time situation regarding lighting requirements is reversed. The luminance level outside the tunnel is then lower than that inside and adaptation problems may occur at the tunnel exit. No difficulties will normally arise so long as the ratio of the luminance level inside the tunnel to that outside is less than 3 : 1. For this reason, exit roads must be provided with an acceptable lighting installation for at least a distance of about 200 to 300 metres from each tunnel exit to aid adaptation (Sec. 5.4). (Where this lighting is to be terminated, this must be done in such a way that the transition from a lighted to an unlit section of road is stepped in level (Sec. 15.2.2).)

17.3 Maintenance

A great deal of dust and grime is brought into a tunnel by the traffic passing through it. In addition, the exhaust fumes of the vehicles contain quantities of unburnt carbon particles. This may reduce visibility in the tunnel and combine with the general pollution to produce soiling of the luminaires, daylight screens and light-reflecting surfaces (e.g. road surface and walls). This soiling can bring about a rapid deterioration in the various luminances and so add further to the worsening of the visibility.

Cleaning needs to be frequent and in line with a layed down maintenance programme, only then will the maintenance factor applied when fixing the luminance levels in the design stage remain valid.

As to the method of cleaning the luminaires, washing them with a mechanical scrubber has been found to give the best results; water-jets are not very effective. Quite a number of tunnels are cleaned by special vehicles designed solely for this purpose. It need hardly be added, that the luminaires should be easy to clean, robust (mechanical scrubbers can be vigorous in operation), and proof against corrosion from soiling. Their water resistance should be as specified by the classification 'jetproof'.

Finally, the tunnel walls should have a surface finish that facilitates cleaning.

Bibliography

Adrian, W., *Neuere Untersuchungen über die Blendung in der Strassenbeleuchtung* (New investigations on glare in road lighting). Electrotechniek, (The Netherlands) vol. 44 p. 510 (1966).

Adrian, W. and Eberbach, K., *Über den Zusammenhang zwischen Sehschwelle und Umfeldgrösse* (About the relation between contrast threshold and dimensions of background). Optik, vol. 28, p. 132 (1968/69).

Adrian, W., *Die Unterschiedsempfindlichkeit des Auges und die Möglichkeit ihrer Berechnung* (The contrast sensitivity of the eye and the possibility of its calculation). Lichttechnik, vol. 21 p. 2A (1969).

Adrian, W. and Schreuder, D. A., *A simple method for the appraisal of glare in street lighting*. Lighting Research and Technology, vol. 2 p. 61 (1970).

Adrian, W., *Grundlagen der physiologischen und psychologischen Blendung und ihre numerische Darstellung* (Fundamentals of physiological and psychological glare and their numerical representation). Symposium Schweiz. Lichttechn. Gesellschaft, Zürich (1975).

Adrian, W., *Method of calculating the required luminances in tunnel entrances*. Lighting Research and Technology, vol. 8 p. 103 (1976).

Adrian, W., *The lighting of tunnel entrances in daytime*. Paper to be presented at CIBS National Lighting Conference, Canterbury (1980).

AFE, *Recommendations relatives à l'éclairage des voies publiques* (Recommendations for the lighting of public roads). l'Association Française de l'Eclairage, 5e edition, (France) (1978).

AIDI, *L'illuminazione delle strade e delle gallerie.* (Lighting of roads and tunnels) (M. Bonomo), Associazione Italiana di Illuminazione (Italy) (1974).

AINSI/IES, *American National Standard practice for roadway lighting.* (ANSI/IES RP-8-1977) (1977).

Balder, J. J. and Schreuder, D. A., *Problems in tunnel lighting*. International Lighting Review, vol. 10 p. 24 (1959).

Balder, J. J. and Meulders, G. J., *Die Messung von Iso Leuchtdichte Kurven in der Strassenbeleuchtung durch eine Fernsehübertragungsanlage* (The measurement of iso-luminance curves in road lighting with a television system). Lichttechnik, vol. 21 p. 102 A (1969).

BCH, *Recommendations for roadlighting in USSR*. Guide BCH-22-75 (1975).

Berg, P. von, *Reflexionsverhalten von Fahrbahnbelägen* (Reflection properties of road surfaces). Technische Hochschule, Aachen, Institut für Strassenwesen. (1978).

Bergmans, J., *Lichtreflectie door Wegdekken* (Light reflection by road surfaces). University of Technology, Delft, thesis (1938).

Blackwell, H. R., *Contrast thresholds of the human eye*. J.O.S.A., vol. 36 p. 624 (1946).

Blackwell, H. R., *Development of procedures and instruments for visual task evaluation*. Illuminating Engineering, vol. 65 p. 267 (1970).

Blackwell, O. M. and Blackwell, H. R., *A proposed procedure for predicting performance aspects of roadway lighting in terms of visibility*. Journal of IES, vol. 6 p. 148 (1977).

Boer, J. B. de, *Fundamental experiments of visibility and admissible glare in road lighting*. CIE XII Session Stockholm, (1951).

Boer, J. B. de, Onate, V., Oostrijck, A., *Practical methods for measuring and calculating the luminance of road surfaces*. Philips Res. Rep., vol. 7 p. 54 (1952).

Boer, J. B. de, *Observations on discomfort glare in street lighting; influence of the colours of light*. CIE XIII Session Zürich, (1955).

Boer, J. B. de, Burghout, F., Heemskerck Veeckens, J. F. T. van; *Appraisal of the quality of public lighting based on road surface luminance and glare*. CIE XIV Session Brussels, P-59-23 (1959).

Boer, J. B. de, *Untersuchungen über den Einfluss der Lichtfarbe auf das Sehen im Strassenverkehr* (Investigations on the influence of colour of light on vision in road traffic). Zentralblatt für Verkehrs-Medizin, Verkehrs Psychologie, vol. 6 (1960).

Boer, J. B. de and Knudsen, B., *The pattern of road luminance in public lighting*. CIE XV Session Vienna, P-63.17 (1963).

Boer, J. B. de and Westermann, H. O., *Caractérisation et classification des revêtements routiers du point de vue de la luminance en éclairage public* (Characterisation and classification of road surfaces from the point of view of luminance in public lighting). Lux, vol. 30 p. 385 (1964a).

Boer, J. B. de and Westermann, H. O., *Die Unterscheidung von Strassenbelägen nach Reflexionskennwerten und deren Bedeutung für die Strassenbeleuchtung* (The distinguishment of road surfaces in dependance of the reflection properties and its meaning for road lighting). Lichttechnik, vol. 16 p. 487 (1964b).

Boer, J. B. de and Schreuder, D.A., *Glare as a criterion for quality in street lighting*. Trans. Ill. Eng. Soc. (London), vol. 32 p. 117 (1967).

Boer, J. B. de, Public Lighting. Philips Technical Library, (1967a).

Boer, J. B. de and Vermeulen, J., *Simple luminance calculations based on road surface classification*. CIE XVI Session Washington, P-67.14 (1967b).

Boer, J. B. de, *Quality aspects in public lighting*. Engineering Report 24, Philips, Eindhoven (1972).

Boer, J. B. de, *Modern light sources for highways*. Journal of IES, vol. 3 p. 142 (1974).

Boereboom, A., *The lighting of motorways.* Electrotechniek (The Hague), vol. 44, pp. 543–548 (1966).

Bommel, W. J. M. van, *Gepolariseerd licht en de toepassing daarvan voor autoverlichting* (Polarised light and its application for vehicle lighting). University of Technology, Eindhoven (1970).

Bommel, W. J. M. van, *Mobile laboratory for outdoor lighting.* International Lighting Review, vol. 27 p. 8 (1976a).

Bommel, W. J. M. van, *Optimisation of the quality of roadway lighting installations, especially under adverse weather conditions.* Journal of IES, vol. 55 p. 99 (1976b).

Bommel, W. J. M. van, *Road lighting for all weathers.* J. Illum. Eng. Inst, (Japan), 60, p. 4 (1976c).

Bommel, W. J. M. van, *Road lighting performance sheets and their use for design and investigation purposes.* Trans. of CIE TC-4.6 Symposium, Karlsruhe (1977).

Bommel, W. J. M. van, *Design considerations for roadway lighting.* Journal of IES, vol. 8 p. 40 (1978a).

Bommel, W. J. M. van, *Optimisation of road lighting installations by the use of performance sheets.* Lighting Research and Technology, vol. 10 p. 189 (1978b).

Bommel, W. J. M. van, *Interrelation of road lighting quality criteria.* Lichtforschung, vol. 1, p. 10 (1979).

Bommel, W. J. M. van, *Tunnel lighting world-wide.* Paper to be presented at CIBS National Lighting Conference, Canterbury (1980).

Borel, P., *Unfallverhütung und öffentliche Beleuchtung* (Avoidance of accidents and public lighting). Bull. Schweiz. Elektrotechn. Ver., vol. 49 p. 8 (1958).

Bouguer, *Traité d'optique* (About-Optics). 1760.

Box, P. C., *Relationship between illumination and freeway accidents.* Illuminating Engineering, vol. 66 p. 365 (1971).

BSI, *Lighting for underpasses and bridged roads.* Code of Practice for Road Lighting, CP 1004: Part 7: 1971, British Standards Institution (1971).

BSI, *General Principles.* Code of Practice for Road Lighting, CP 1004: Part 1: 1973, British Standards Institution (1973).

BSI, *Lighting for traffic routes* (Group A). Code of Practice for Road Lighting, CP 1004: Part 2: 1974, British Standards Institution (1974).

BSI, *Specification Electric Luminaires.* Part 2 Detail requirements Section 2.7 Road lighting lanterns, BS 4533: Part 2.7: 1976, British Standards Institution (1976).

Buck, J. A., McGowan, T. K. and McNelis, J. F., *Roadway visibility as a function of light source colour.* Journal of IES, vol. 5 p. 20 (1975).

Burghout, F., *Kenngrössen der Reflexionseigenschaften von trockenen Fahrbahndecken* (Characteristic figures for the reflection properties of dry road surfaces). Lichttechnik, vol. 29, p. 23 (1977a).

Burghout, F., *Simple parameters significant of the reflection properties of dry road surfaces.* Trans. of CIE TC-4.6 Symposium, Karlsruhe, (1977b).

Burghout, F., *On the relationship between reflection properties, composition and texture of road surfaces.* CIE XIX Session Kyoto, P-79.65 (1979).

Bijllaardt, D. van den, Fraipont, W. J. de, Tan, T. H. and Westermann, H. O., *A comparison between catenary and conventional, pole mounted, lighting systems for motorways.* CIE XVIII Session London, P-75.59 (1975).

Bijllaardt, D. van den, *De invloed van de armatuuropstelling op de gelijkmatigheid van de wegdekluminantie* (The influence of the luminaire arrangement on the uniformity of the road surface luminance). Elektrotechniek (The Netherlands) vol. 54, p. 290 (1976).

Bijllaardt, D. van den, *Praktische Aspekte der Tunnelbeleuchtung* (Practical aspects of tunnel lighting). Lichttechnik vol. 29, p. 61, (1977).

Bijllaardt, D. van den, *Basic points of tunnel lighting.* To be published (1980).

Caminada, J. F., *De functies van straatverlichting in woonwijken, les fonctions de l'éclairage public dans les zones residentielles* (The functions of street lighting in residential areas). Bulletin Société Royale Belge des Electriciens, Koninklijke Belgische Vereniging der Electrotechnici, vol 95 p. 71, (1979).

Caminada, J. F. and Bommel, W. J. M. van, *New lighting criteria for residential areas.* Paper to be presented at National Lighting Conference of IES, Dallas (1980).

Campbell, F. W. and Gubisch, R. W., *The effect of chromatic abberation on visual acuity.* J. Physiol. vol. 192 p. 345 (1967).

Christie, A. W., *An experimental low-cost lighting system for rural highways.* Light and Lighting, vol. 55 p. 270 (1962).

Christie, A. W., *Proceedings of the Institution of Municipal Engineers.* Annual Conference, Blackpool (1966).

CIE, *International recommendations for the lighting of public thoroughfares.* CIE Publication No. 12 (first ed.), (1965).

CIE, *A unified framework of methods for evaluating visual performance aspects of lighting.* CIE Publication No. 19 (TC-3.1), (1972).

CIE, *Photometry of luminaires for street lighting.* CIE Publication No. 27 (TC-2.4), (1973a).

CIE, *Recommendations for tunnel lighting.* CIE Publication No. 26 (TC-4.6), (1973b).

CIE, *Calculation and measurement of luminance and illuminance in road lighting.* CIE Publication No. 30 (TC-4.6), (1976a).

CIE, *Glare and uniformity in road lighting installations.* CIE Publication No. 31 (TC-4.6), (1976b).

CIE, *Recommendations for the lighting of roads for motorized traffic.* CIE Publication No. 12 (sec. ed.) (TC-4.6), (1977a).

CIE, *Depreciation of installations and their maintenance.* CIE Publication No. 33 (TC-4.6), (1977b).

CIE, *Guide on the Emergency Lighting of Premises.* (TC-4.1) draft (1979a).

CIE, *Luminance in the threshold zone.* (TC-4.6), draft (1979b).

290

CIE, *Street lighting and accidents*. CIE Publication No. 8/2 (TC-4.6), draft (1979c).

CIE, *Road lighting for wet conditions*. CIE Publication No. 47 (TC-4.6), (1980).

Claxton, M. J., *The Humberside story*. International Lighting Review, vol. 4 p. 110, (1978).

Cobb, J., Hargroves, R. A., Marsden, A. M. and Scott, P. P., *Road lighting and accidents*. CIE XIX Session Kyoto, P-79.63, (1979).

Cornwell, P., *A study of lighting and road traffic*. University of Birmingham, thesis (1971).

Cornwell, P. R. and Mackay, G. M., *Public lighting and road accidents*. Traffic Engineering and Control, vol. 14 p. 142 (1972).

Crawford, B. H., *The integration of the effects from a number of glare sources*. Proc. Phys. Soc. London, vol. 48 p. 35 (1936).

DEN, *Projektering Vejbelysning (9.20)*, *Vcj regler vedrøprende projektering av vej-belysningsanlegg*. (The design of road lighting; rules with regard to the design of road lighting). Vejdirektoratet Vejregelsekretaratet, (Denmark) draft (1976).

DEN, *Vejbelysning* (Road lighting). 9.20.0/9.20.02 Udstyr (Denmark), (1979).

DIN, *Beleuchtung von Strassentunneln und Unterführungen* (The lighting of road tunnels and underpasses). Vornorm, DIN 67524 (Germany), (1972).

DIN, *Beleuchtung von Strassen für den Kraftfahrzeugverkehr* (Lighting of roads for motorised traffic). Entwurf, DIN 5044, Teil 2 (draft), (1979).

Economopoulos, I. A., *Relationship between lighting parameters and visual performance in road lighting*. Trans. of CIE TC-4.6 Symposium, Karlsruhe (1977).

Economopoulos, I. A., *Photometric parameters and visual performance in road lighting*. University of Technology, Eindhoven, thesis (1978).

Epaneshnikov, M. M., Obrosova, N. A., Sidorova, T. N. and Undasynov, G. N., *New characteristics of lighting conditions for premises of public buildings and methods for their calculation*. CIE XVII Session Barcelone, P-71.30, (1971).

Erbay, A., *Verfahren zur Kennzeichnung der Reflexionseigenschaften von Fahrbahndecken* (Method for the characterisation of the reflection properties of road surfaces). University of technology, Berlin, thesis (1973).

Erbay, A., *Der Reflexionseigenschaften von Fahrbahndecken* (The reflection properties of road surfaces). University of technology, Berlin, (1974a).

Erbay, A., *Ein neues Verfahren zur Kennzeichnung der Reflexionseigenschaften von Fahrbahndecken* (A new method for the characterisation of the reflection properties of road surfaces). Lichttechnik, vol. 26 p. 239 (1974b).

Ferguson, H. M. and Stevens, W. R., *Relative brightness of coloured light sources*. Trans. of the Ill. Engn. Soc. (London), vol. 21 p. 227 (1956).

FfdS, *Richtlinien für die Beleuchtung in Anlagen für Fussgängerverkehr* (Guidelines for the lighting of pedestrian areas). Forschungsgesellschaft für das Strassenwesen, Köln, (1977).

Fischer, D., Kebschull, W., *Kostenvorteile von Niederdruck-Natriumdampflampen in der Strassenbeleuchtung* (Cost advantages of low pressure sodium lamps in road lighting). Lichttechnik, nr. 12 p. 520 (1978a).

Fischer, D., *Security lighting*. International Lighting Review, vol. 4 p. 127 (1978b).

Fischer, D., *Pedestrian areas*. International Lighting Review, vol. 4 p. 17 (1978c).

Fisher, A. J. and Christie, A. W., *A note on disability glare*. Vision Research, vol. 5 p. 565 (1965).

Fisher, A. J. and Hall, R. R., *Road luminances based on detection of change of visual angle*. Lighting Research and Technology, vol. 8 p. 187 (1976).

Fisher, A. J., *Road lighting as an accident counter measure*. Trans. of CIE TC-4.6 Symposium, Karlsruhe (1977).

Foitzik, L., Zschaeck, H., *Messungen der spektralen Zerstreuungsfunktion bodennaher Luft bei guter Sicht, Dunst und Nebel* (Measurements of the spectral scatteringsfunction of sky at low altitudes with good visibility, light fog and dense fog). Zeitschrift für Meteorologie, vol. 7, p. 1, (1953).

Frederiksen, E., *Uni-directional-Sensitive Photometer*. Light and Lighting, vol. 60 p. 46 (1967).

Frederiksen, E., *Intercomparison test on the accuracy of photometric measurements of road surfaces*. CIE XVII Session Barcelone, P-71.06, (1971).

Frederiksen, E. and Gudum, J., *The quality of street lighting under changing weather conditions*. Lighting Research and Technology, vol. 4 p. 90 (1972).

Frederiksen, E. and Sørensen, K., *Reflection classification of dry and wet road surfaces*. Lighting Research and Technology, vol. 8 p. 175 (1976).

Fritz, H. and Pusch, R., *Das Strassenreflektometer* (The road surface reflectometer). Siemens Elektrodienst, vol. 11, p. 8 (1969).

Fry, G. A., Pritchard, B. S. and Blackwell, H. R., *Design and calibration of a disability glare lens*. Illuminating Engineering, p. 120, (1963).

Fujimori, S., *Paper presented at the annual session of the Japanese Highway Public Corporation (1973)*.

Gallagher, V. P. and Meguire, P. G., *Contrast requirements of urban driving*. Transportation Research Board Special Report 156 (Washington) (1975).

Gallagher, V. P., *A visibility metric for safe lighting of city streets*. Journal of IES, vol. 5 p. 85 (1976).

Grijs, J. C. de, *Visuele beoordeling van verlichtingscriteria in Den Haag en Amsterdam* (Visual assessment of lighting criteria in The Hague and Amsterdam). Elektrotechniek (The Netherlands), vol. 50 p. 19 (1972).

Harris, A. J. and Christie, M. A., *The revealing power of street lighting installations and its calculation*. Trans. Illum. Eng. Soc. (London) vol. 16 p. 120 (1951).

Hartmann, E. and Moser, A. E., *Das Gesetz der physiologischen Blendung bei sehr kleinen Blendwinkeln* (The rule of physiological glare for very small glare angles). Lichttechnik, vol. 20 p. 67 A (1968).

Hentschel, H. J., *A physiological appraisal of the revealing power of street lighting installations for large composite objects.* Lighting Research and Technology, vol. 3 p. 268 (1971).

Holladay, L. L., *Action of a light source in the field of view in lowering visibility.* J.O.S.A., vol. 14 p. 1 (1927).

Hopkinson, R. G., *Discomfort glare in lighted street.* Trans. Ill. Eng. Soc. (London), vol. 5 p. 1 (1940).

HRFJ, *Recommendations for the lighting of vehicular traffic tunnels.* Express Highway Research Foundation of Japan, (1979).

IES, *Lighting of tunnels.* Journal of IES, vol. 1 p. 247, (1972).

ITE/IES, *Public lighting needs.* Illuminating Engineering, vol. 61 p. 585 (1966).

Jackett, M. J., Fisher, A. J., *The light reflecting properties of asphaltic concrete road surfaces.* Australian Road Research Board (ARRB) Proceedings, vol. 7, part. 5, (1974).

Jainski, P., *Die Sehschärfe des menschlichen Auges bei verschiedenen Lichtfarben* (The visual acuity of the human eye for different colours of light). Lichttechnik, vol. 12 p. 402 (1960a).

Jainski, P., *Die Unterschiedsempfindlichkeit des menschlichen Auges bei verschiedenen Lichtfarben* (The contrast sensitivity of the human eye under different colours of light). Lichttechnik, vol. 12 p. 35 (1960b).

Jainski, P., *Das Verhalten der Unterschiedsempfindlichkeit bei Blendung mit verschiedenen Lichtfarben* (The behaviour of the discrimination sensitivity with glare from different kinds of light). Lichttechnik, vol. 14 p. 60 (1962).

Janoff, M. S., Koth, B., McCunny, W., Freedman, M., Duerk, C. and Berkovitz, M., *Effectiveness of highway arterial lighting.* Federal Highway Administration Report No. FHWA-RD-77-33, Washington (1977).

Janoff, M. S., Koth, B., McCunny, W., Berkovitz, M. J. and Freedman, M., *The relationship between visibility and traffic accidents.* Journal of IES, vol. 8 p. 95 (1978).

Jantzen, R., *Flimmerwirkung der Verkehrsbeleuchtung* (Flicker in traffic lighting). Lichttechnik, vol. 12 p. 211 (1960).

JIS, *Lighting of Traffic Routes.* Japanese Industrial Standards (JIS-Z 9111-1969) (1969).

JSR, Japanese Standard on Road Lighting (JSR). Ministry of construction, road bureau (1967).

Kabayama, H., *Study on adaptive illumination for sudden change of brightness.* J. Illum. Eng. Inst. (Japan) vol. 47 p. 488 (1963).

Kebschull, W., *Die Reflexion trockener und feuchter Strassenbeläge* (The reflection of dry and moist road surfaces). University of technology, Berlin, thesis (1968).

Ketvertis, A., *Road illumination and traffic safety.* Road and Motor vehicle traffic safety branch of Transport, Ottawa (1977).

Ketvertis, A., *Integration of luminance and illuminance calculation method in North American Roadway lighting design*. APLE/CIE TC-4.6 Symposium, Manchester (1979).

Lambert, G. K., Marsden, A. M. and Simons, R. H., *Lantern intensity distribution and installation performance*. Public Lighting, vol. 38 p. 27 (1973).

Languasco, F., Pasta, M. and Soardo, P., *Photometry of luminaires at I.E.N.G.F.: a new goniophotometer*. CIE XIX Session Kyoto, P-79.30 (1979).

LiTG, *Strassenreflektometer Messvergleiche* (Road surface reflectometer comparisons). The German Illuminating Engineering Society (to be published).

LTAG, *Oesterreichische Richtlinien für Strassenbeleuchtungsanlagen* (Austrian recommendations for road lighting installations). Oesterreichischen Lichttechnischen Arbeitsgemeinschaft (LTAG) (1979).

Luckiesh, M. and Moss, F. K., *Visual acuity and sodium vapour light*. Journal of the Franklin Institute, vol. 215 p. 401 (1933).

Luckiesh, M. and Moss, F. K., *Seeing in tungsten, mercury and sodium light*. Trans. Ill. Eng. Soc. (America), vol. 31 p. 655 (1936).

Mäder, F. and Fuchs, O., *Beitrag zur Frage der Eingangsbeleuchtung von Strassentunneln* (Contribution with regard to entrance lighting of road tunnels). Bull. SEV (Switzerland), vol. 57 p. 359 (1966).

Mäder, F., *Verwendung von besonderen asymmetrischen Leuchten* (Schrägstrahlern) *zur Beleuchtung von Tunneln* (Application of special assymetrical luminaires for the lighting of tunnels). Bull. SEV (Switzerland), vol 60, p. 1177 (1969).

Massart, P., *Définition d'un revêtement de synthese en éclairage routier* (Definition of synthetical surfaces in road lighting). University of Liège, thesis (1973).

MOP, *Alumbrado de carreteras* (Lighting of roads). Direccion General de Carreteras y Caminos Vecinales (Spain), (1964).

Morass, W. and Rendl, F., *Portable measuring instrument for road-surface luminances*. Light and Lighting, vol. 60, p. 157, (1967).

MSZ, *Közvilagitas letesitese* (Construction of public lighting). MSZ 09-0214-74, (Hungary), (1976).

Müller, E. and Riemenschneider, W., *Tunnel lighting in Switzerland*. Lighting Research and Technology, vol. 7 no. 2 p. 99, (1975).

Narisada, K., *Influence of non-uniformity in road surface luminance of public lighting installations upon perception of objects on the road surface by car-drivers*. CIE XVIII Session Barcelone, P-71.17 (1971).

Narisada, K. and Yoshikawa, K., *Tunnel entrance lighting-effect of fixation point and other factors on the determination of requirements*. Lighting Research and Technology, vol. 6 p. 9 (1974).

Narisada, K. and Yoshimura, Y., *National Technical Report*. Vol. 20 p. 287, Japan (1974).

Narisada, K., *Applied Research on tunnel entrance lighting in Japan*. Lighting Research and Technology, vol. 7 p. 87 (1975).

Narisada, K. and Yoshikawa, K., *Lighting of short tunnels, effect of the exit luminance on the level of lighting*. CIE XVIII Session London, P-75.58 (1975).

Narisada, K., Yoshikawa, K. and Yoshimura, Y., *Adaptation luminance of driver's eyes approaching a tunnel entrance in daytime*. CIE XIX Session Kyoto, P-79.64, (1979).

NBN, *Leidraad voor verlichting van de openbare wegen* (Guide for the lighting of public roads). Belgisch instituut voor normalisatie (Norm NBN-L19.001), (Belgium), (1979).

NBR, *Road Design Specifications*. Part: Road Lighting. National Board of Public Roads and Waterways, draft, (Finland), (1979).

NRA, *Riktlinjer foer stationaer vaegbelysning* (Guidelines for stationary road lighting). National Road Administration, draft, (Sweden), (1975).

NSvV, *Aanbevelingen voor tunnelverlichting* (Recommendations on tunnel lighting). Nederlandse Stichting voor Verlichtingskunde (The Netherlands), (1963).

NSvV, *Richtlijnen en Aanbevelingen voor openbare verlichting* (Guidelines and recommendations for public lighting). Nederlandse Stichting voor Verlichtingskunde, (The Netherlands), (1974).

Øbro, P. and Sørensen, K., *A portable instrument for the measurement of road reflection properties*. Internal report, The Danish Illuminating Engineering Laboratory (1976).

PKNM, *Oswietlenie dróg publicznych* (Roadway lighting). Polski Komitet Normalizacji i Miar (PKNM) Poland, (1976).

Prevot, M., *Incidence de la nature des revêtements sur la qualité et les coûts des installations d'éclairage public* (Influence of the road surface properties on the quality and costs of public lighting installations). Lux, p. 123 (1978).

Range, H. D., *Ein vereinfachtes Verfahren zur Lichttechnische Kennzeichnung von Fahrbahnbelägen* (A simplified method for the characterisation of road surfaces from a lighting point of view). Lichttechnik, vol. 24, p. 608 (1972).

Range, H. D., *Strassenreflektometer zur vereinfachten Bestimmung Lichttechnische Eigenschaften von Fahrbahnbelägen* (Road reflectometer for the simplified determination of the properties of lighted road surfaces). Lichttechnik, vol. 25, p. 389 (1973).

Roch, J. and Smiatek, G., *Das q_s-Verfahren (nach Roch/Smiatek) zur lichttechnischen Kennzeichnung von Fahrbahndecken* (The q_s-method for the characterisation of road surfaces for lighting). Lichttechnik, vol. 24 p. 329 (1972).

Roch, J., *Auswirkung einer repräsentativen Untersuchung von Fahrbahnbelägen auf die Planung von Strassenbeleuchtungsanlagen* (The use of a representative investigation of road surfaces for the design of road lighting installations). Factagung der LiTG, SLG and LTAG, Salzburg, (1974).

RRL, *Research on road safety*. D.S.I.R. Road Research Laboratory H.M.S.O. London, (1963).

SAA, *Public Lighting Code*. Part 1, Lighting of urban traffic routes. Australian Standard 1158, Part 1 – (1973).

Sabey, B. E., *Road surface reflection characteristics*. Transport and Road Research Laboratory Report LR490, (1971).

Sabey, B. E. and Johnson, H. D., *Road lighting and accidents: before and after studies on trunk road sites.* Transport and Road Research Laboratory Report, LR 586 (1973).

Sabey, B. E., *Potential for accident and injuring reduction in road accidents.* Traffic Safety seminar, Road Traffic Safety Research Council of New Zealand, (1976).

Sarteel, F., *Les implications économiques du contrôle de d'éblouissement en éclairage public* (The economic consequences of the control of glare in public lighting). CIE XVIII Session London, P-75.55 (1975).

Scholz, I., *Street lighting and motoring accidents.* International Lighting Review, vol. 4 p. 128, (1978).

Schreuder, D. A., *The lighting of vehicular traffic tunnels.* University of Technology, Eindhoven, thesis (1964).

Schreuder, D. A., *Tunnel lighting.* Chapter 4 of Public Lighting (edited by J. B. de Boer), (1967).

Schreuder, D. A., *Ein Vergleich von Empfehlungen für Tunneleinfahrt-Beleuchtung* (A comparison of recommendations on tunnel entrance lighting). Lichttechnik, vol. 20 p. 20 A, (1968).

Schreuder, D. A., *De bepaling van verblinding bij openbare verlichting* (The determination of glare in public lighting). Elektrotechniek (The Netherlands), vol. 50 p. 26, (1972).

Schröter, H. G., *Research on daytime lighting of tunnel entrances.* Lighting Research and Technology, vol. 9 p. 194, (1977).

SEV, *Leisätze für öffentliche Beleuchtung* (Guide for public lighting). 2. Teil, Schweizerischer Elektroteknischer Verein, SEV 4024 (Switzerland), (1968).

SEV, *Leitsätze der Schweizerischen Lichttechnische Gesellschaft Offentliche Beleuchtung* (Guide of the Swiss lighting society on public lighting). Schweizerischer Elektrotechnische Verein (SEV), 1. Teil: Allgemeine Richtlinien SEV 8907-1.1977, (Switzerland) (1977).

Skene, P., *The cost effectiveness of upgrading urban street lighting.* School of Transportation and Traffic (1976).

Smith, F. C., *Reflection factors and revealing power.* Trans. Illum. Eng. Soc. (London), vol. 3 p. 196 (1938).

Sørensen, K., *Description and classification of light reflection properties of road surfaces.* The Danish Illuminating Engineering Laboratory, Report No. 7, (1974).

Sørensen, K., *Road surface reflection data.* The Danish Illuminating Engineering Laboratory, Report No. 10, (1975).

Spinelli, J. A., *Low pressure sodium lamps solve streetlighting problems.* Electrical Construction and Maintenance, p. 53, (1976).

Spinelli, J. A., *New street lighting in Long Beach.* International Lighting Review, vol. 4 p. 103, (1978).

STLR, *Vögbeläggningers Lystekniske Egenskaber* (Lighting properties of road surfaces). Scandinavian Traffic Lighting Research, Report No. 2, 1978.

Turner, H. J., *The effectiveness of the New South Wales Street lighting Subsidy Scheme.* National Road Safety Symposium, Department of Transport, Canterra (1972).

Verbeek, T. G. and Vermeulen, J., *Laboratory method for measuring luminance factors of road surfaces.* Light and Lighting, vol. 64 p. 131, (1971).

Vermeulen, J., *Light reflection from rough surfaces.* CIE XVIII Session London, P-75.80, (1975).

VSV, *Geometrisk utforming – Belysning, generalt* (Geometric layout–lighting, general). Vegnormaler Statens Vegvesen (VSV), AR 1975, Norway (1975).

Waldram, J. H., *The revealing power of street lighting installations.* Trans. Illum. Eng. Soc. (London) vol. 3 p. 173 (1938).

Walthert, R., *Zur Bewertung der Leuchtdichteverteilung beleuchterter Strassen* (The appraisal of the luminance distribution of lighted roads). University of Karlsruhe (thesis), (1973).

Walthert, R., *The influence of lantern arrangement and road surface luminance on subjective appraisal and visual performance in street lighting.* CIE XVIII Session London, P-75.60, (1975).

Walthert, R., *Lichtmessung auf Strassen und in Sportanlagen* (Light measurement of roads and sports lighting installations). Bericht SLG-Tagung Lichtmesstechnik, (Switzerland), (1977a).

Walthert, R., *Tunnel lighting systems.* International Lighting Review, vol. 4 p. 112 (1977b).

Weigel, R. G., *Untersuchungen über die Sehfähigkeit im Natrium- und Quecksilberlicht insbesondere bei der Strassenbeleuchtung* (Investigations on visibility possibilities in sodium and mercury light, especially in road lighting). Das Licht, vol. 5 p. 211 (1935).

Westermann, H. O., *Reflexionskennwerte von Strassenbelägen* (Characteristic reflection figures for road surfaces). Lichttechnik, vol. 15 p. 506 (1963).

Westermann, H. O., *Engineering aspects of tunnel lighting.* Lighting Research and Technology, vol. 7 p. 90 (1975a).

Westermann, H. O., *Vergelijking van de lichttechnische eigenschappen van lijnverlichting en conventionele verlichting* (Comparison of lighting properties of catenary and conventional lighting). Elektrotechniek (The Netherlands) 53, p. 889, (1975b).

Ziegler, W., *Ein Gonio-Reflektometer zur Messung von Strassenbelagsproben* (A gonio-reflectometer for the measurement of road surfaces). Lichttechnik, vol. 30, p. 172 (1978).

Glossary

Adaptation: 1. The process by which the properties of the visual system are modified according to the luminances or the colour stimuli presented to it.
2. The final state of the process.

Adaptation luminance: the luminance value that corresponds to the momentary adaptation state of the eye. (See also **Equivalent adaptation luminance**).

After image: a visual response that occurs after the stimulus causing it has ceased.

Angle of tilt: see **Tilt, angle of.**

Average luminance coefficient (Q_o): a measure for the lightness of a road surface being defined as the value of the luminance coefficient q averaged over a specified solid angle of light incidence.

Ballast: device used with discharge lamps for stabilizing the current in the discharge.

Beam axis: the direction in the centre of the solid angle which is bounded by directions having luminous intensities of 90 % of the maximum intensity of a luminaire.

Beam efficiency: the ratio of the flux emitted within the solid angle defined by the beam spread, to the bare lamp flux.

Beam spread: the angle (in a plane through the beam axis) over which the luminous intensity drops to a stated percentage of its peak intensity.

Black body (Planckian radiator): thermal radiator that absorbs completely all incident radiation, whatever the wavelength, the direction of incidence or the polarization. This radiator has, for any wavelength, the maximum spectral concentration of radiant exitence at a given temperature.

Black body locus (Planckian locus): the line in a chromaticity diagram representing the chromaticity of full (or Planckian) radiators of different temperatures.

Blended-light lamp: lamp containing in the same bulb a high-pressure mercury vapour discharge tube and an incandescent lamp filament connected in series. The bulb may be diffusing or coated with a fluorescent material. For example, the MLL lamp.

Brightness (or luminosity): attribute of visual sensation according to which an area appears to emit or reflect more or less light.
Note: Brightness according to the definition is also an attribute of colour. In British recommendations the term 'brightness' is now reserved to describe brightness of colour; luminosity should be used in all other instances.

Catenary arrangement: arrangement of luminaires suspended with their main beam axes at right angles to the road axis.

Catenary luminaire: luminaire designed to be suspended from a cable with its main beam axis at right angles to the axis of the road.

Central arrangement: see **Twin-central arrangement.**

CIE standard photometric observer: receptor of radiation whose relative spectral sensitivity curve conforms to the $V(\lambda)$ curve or to the $V'(\lambda)$ curve.

Colour appearance: general expression for the colour impression received when looking at a light source.

Colour rendering: general expression for the effect of an illuminant on the colour appearance of objects in conscious or subconscious comparison with their colour appearance under a reference illuminant.

Colour rendering index (R_a): of a light source. Measure of the degree to which the psycho-physical colours of objects illuminated by the source conform to those of the same objects illuminated by a reference illuminant for specified conditions.

Colour temperature: temperature of the black body that emits radiation of the same chromaticity as the radiation considered.
Unit: Kelvin, K.

Cone: retinal receptor element which is presumed to be primarily concerned with perception of light and colour stimuli when vision is photopic.

Contrast: subjective assessment of the difference in appearance of two parts of a field of view seen simultaneously or successively. (See also **Luminance contrast**).

Contrast threshold: see **Threshold contrast.**

Control: luminaire characteristic, determined by the value of the specific luminaire index (SLI), that indicates the degree of glare control present. Luminaires are classified as being of limited, moderate, or tight control.

Corrected specular factor $(S1')$: a factor representing the degree of specular reflection of a wet road surface.

Correlated colour temperature: the colour temperature corresponding to the point on the Planckian locus that is nearest to the point representing the chromaticity of the illuminant considered on an agreed uniform-chromaticity-scale diagram.
Unit: Kelvin, K.

Cosine law of incidence: the law that states that illuminance at a point on a plane is proportional to the cosine of the angle of light incidence (the angle between the direction of the incident light and the normal to the plane).

$$E = \frac{I}{d^2} \cos \gamma$$

Cross factor (CF): a luminaire characteristic that indicates the suitability of the luminaire for use under wet and foggy weather conditions.

Depreciation factor (deprecated): the reciprocal of the maintenance factor.

Detection: recognition of the presence of something without necessarily identifying it.

Diffuse reflection: diffusion by reflection in which, on the macroscopic scale, there is no regular reflection.

Diffuse transmission: transmission in which, on the macroscopic scale, diffusion occurs independently of the laws of refraction.

Disability glare: glare that impairs the vision of objects without necessarily causing discomfort.

Discharge lamp: lamp in which the light is produced, either directly or by means of phosphors, by an electric discharge through a gas, a metal vapour, or mixture of several gases and vapours.

Discomfort glare: glare that causes discomfort without necessarily impairing the vision of objects.

Downward light output ratio: see **Upward light output ratio.**

Driver stopping distance: the total distance travelled while a vehicle is being brought to rest, measured from the position of the vehicle at the instant the driver has an opportunity to perceive that he should stop his vehicle.

Effective (road) width: the horizontal distance between the vertical through a luminaire centre and the kerb farthest from that luminaire.

Efficacy: see **Luminous efficacy.**

Equivalent adaptation luminance: that value of uniform luminance in front of an observer that would result in the same degree of perceptibility as with the actual prevailing non-uniform luminance distribution.

Equivalent veiling luminance: luminance which has to be added, by superposition, to the luminance of both the adapting background and the object in order to make the luminance difference threshold in the absence of disability glare the same as that experienced in the presence of disability glare.

Flashed area: the area of the orthogonal projection of the light emitting surface of a luminaire on a plane perpendicular to a given direction of viewing within which the luminance exceeds 1 % of the luminance of the brightest part. For the purpose of the determination of the glare control mark the direction of viewing is specified as $\gamma = 76°$ in the plane $C = 0°$.

Flicker: impression of fluctuating luminance or colour.

Floodlight: projector designed for floodlighting, usually capable of being pointed in any direction and of weatherproof construction.

Fluorescence: photoluminescence that persists for an extremely short time after excitation.

Fluorescent lamp: discharge lamp in which most of the light is emitted by a layer of fluorescent material excited by the ultraviolet radiation from the discharge.
Note: This term is most commonly applied to low-pressure tubular fluorescent lamps, for example: 'TL', 'TL'F, 'TL'E lamps.

Glare: condition of vision in which there is discomfort or a reduction in the ability to see significant objects, or both, due to an unsuitable distribution or range of luminance or to extreme contrasts in space or time.

Glare control mark: number denoting the degree to which discomfort glare is controlled.

Halide lamp: see **Metal halide lamp.**

Halogen lamp: gas-filled lamp containing a tungsten filament and a small proportion of halogens.

High-pressure mercury (vapour) lamp: mercury vapour lamp, with or without a coating of phosphor, in which during operation the partial pressure of the vapour is of the order of 10^5 Pa – for example: HPL and HPL-N lamps.

High-pressure sodium (vapour) lamp: sodium vapour lamp in which the partial pressure of the vapour during operation is of the order of 10^4 Pa – for example: SON and SON/T lamps.

Ignitor: see **Starter.**

Illuminance *(E)*: at a point on a surface. Quotient of the luminous flux incident on an element of the surface containing the point and the area of that element.
Unit: lux, lx.
E_h: illuminance on a horizontal surface.
E_v: illuminance on a vertical surface.
E_c: illuminance on a cylindrical surface.
E_s: illuminance on a spherical surface.

Illuminance meter: an instrument for the measurement of illuminance.

Illumination: the application of visible radiation to an object.

Incandescence: emission of visible radiation by thermal excitation.

Incandescent (electric) lamp: lamp in which light is produced by means of a body heated to incandescence by the passage of an electric current.

Inclination, angle of: see **Tilt, angle of.**

Inter-reflection (or interflection): general effect of the reflections of radiation between several reflecting surfaces.

Intensity *(I)*: see **Luminous intensity.**

Intensity distribution: see **Luminous intensity distribution.**

Inverse Square Law: the law that states that the illuminance at a point on a plane perpendicular to the line joining the point and a source is inversely proportional to the square of the distance between the source and the plane.

$$E = \frac{I}{d^2}$$

Isocandela curve (diagram): curve traced on an imaginary sphere with the source at its centre and joining all adjacent points corresponding to those directions in which the luminous intensity is the same, or a plane projection of this curve.

Isoluminance curve (diagram): locus of points on a surface at which the luminance is the same, for given positions of the observer and of the source or sources in relation to the surface.

Isolux curve (diagram): locus of points on a surface where the illuminance has the same value.

Kerb ratio: the ratio of the average illuminance on the footpath of a specified width to the average illuminance on the same width of adjacent carriageway.

Lamp: source made in order to produce light.

Lamp mortality: see **Mortality rate.**

Landolt ring: two dimensional ring with a gap, the width of the gap and the thickness of the ring each being equal to $^1/_5$ of the ring's outer diameter.

Lantern: see **Luminaire.**

Light: any radiation capable of causing a visual sensation directly i.e. **Visible radiation.**

Light distribution: see **Luminous intensity distribution.**

Light output ratio: of a luminaire, is the ratio of the light output of the luminaire, measured under specified practical conditions, to the sum of the individual light outputs of the lamps operating outside the luminaire under specified conditions.

Lightness: attribute of visual sensation in accordance with which a body seems to transmit or reflect diffusely a greater or smaller fraction of the incident light.

Longitudinal uniformity (U_l): the ratio of minimum to maximum luminance along a line parallel to the road axis through the observer's position.

Louvre: screen made of translucent or opaque components and geometricelly disposed to prevent lamps from being directly visible over a given angle.

Low-pressure mercury (vapour) lamp: mercury vapour lamp, with or without a coating of phosphor, in which during operation the partial pressure of the vapour does not exceed 100 Pa – for example: a 'TL' lamp.

Lower luminous flux: see **Upper luminous flux.**

Low-pressure sodium lamp: sodium vapour lamp in which the partial pressure of the vapour during operation does not exceed 5 Pa – for example: a SOX lamp.

Luminaire: apparatus that distributes, filters or transforms the lighting given by a lamp or lamps and which includes all the items necessary for fixing and protecting these lamps and for connecting them to the supply circuit.
Note: In road lighting the term 'lantern' is also sometimes used.

Luminance (L): (in a given direction, at a point on the surface of a source or a receptor or at a point on the path of a beam). Quotient of the luminous flux leaving, arriving at, or passing through an element of surface at this point and propagated in directions defined by an elementary cone containing the given direction and the product of the solid angle of the cone and the area of the orthogonal projection of the element of surface on a plane perpendicular to the given direction.
Unit: candela per square metre, cd/m^2.

Luminance contrast (C): between two parts of a visual field, is the relative luminance difference of those parts in accordance with the formula:

$$C = \frac{|L_1 - L_2|}{L_2}$$

where the size of the two parts differs greatly and where
L_1 = luminance of the smallest part (the object),
L_2 = luminance of the greatest part (the background).

302

Luminance coefficient (q): the ratio, for a specified direction of observation and direction of light incidence, between the luminance on an element of a given road surface and the illuminance on it.
Unit: candela per square metre per lux, $cd/m^2/lux$.

Luminance meter: an instrument for the measurement of luminance.

Luminance yield factor: ratio of the average luminance (in cd/m^2) to the average illuminance (in lux) of a road lighting installation.

Luminosity: see **Brightness.**

Luminous efficacy (of a source): quotient of the luminous flux emitted and the power consumed.
Unit: lumen per watt. lm/W.

Luminous flux (Φ_v), (Φ): the quantity derived from radiant flux by evaluating the radiation according to its action upon a selective receptor, the spectral sensitivity of which is defined by the standard spectral luminous efficiencies.
Unit: lumen, lm.

Luminous intensity (I_v, I): of a source in a given direction. Quotient of the luminous flux leaving the source, propagated in an element of solid angle containing the given direction, and the element of solid angle.
Unit: candela, cd.
Note: The luminous intensity of luminaires is normally given either in a **Luminous intensity** diagram or in an **Isocandela diagram.**

Luminous intensity diagram (table): luminous intensity shown in the form of a polar diagram or table, in terms of candela per 1000 lumens of lamp flux. The diagram (table) for non-symmetrical light distributions gives the light distribution of a luminaire in at least two planes:
1. In a vertical plane through the longitudinal axis of the luminaire.
2. In a plane at right angles to that axis.
Note: The luminous intensity diagram (table) can be used:
a. To provide a rough idea of the light distribution of the luminaire.
b. For the calculation of illuminance values at a point.
c. For the calculation of the luminance distribution of the luminaire.

Luminous intensity distribution: distribution of the luminous intensities of a lamp or luminaire in all spatial directions.

Lux meter: see **Illuminance meter.**

Maintenance factor: ratio of the average illuminance on the working plane after a specified period of use of a lighting installation to the average illuminance obtained under the same conditions for a new installation.
Note: The use of the term Depreciation factor as the reciprocal of maintenance factor is deprecated.

Mesopic vision: vision intermediate between photopic and scotopic vision.

Metal halide lamp: discharge lamp in which the light is produced by the radiation from a mixture of a metallic vapour (for example, mercury) and the products of the dissociation of halides (for example, halides of thallium, indium or sodium) – for example: HPI/T lamps.

Monochromatic radiation: radiation characterized by a single frequency or wavelength. By extension, radiation of a very small range of frequencies or wavelengths which can be described by stating a single frequency or wavelength.

Mortality rate: the number of operating hours elapsed before a certain percentage of the lamps fail.

Mounting height: the vertical distance between road surface and centre of the light source.

Opposite arrangement: an arrangement in which the luminaires are placed on either side of the carriageway(s) opposite to one another.

Overall uniformity (U_o): the ratio of minimum to the average luminance over the area of road considered.

Overhang: the horizontal distance between a vertical line passing through the luminaire centre and the nearest kerb of the road.

Peak intensity: the luminous intensity of a luminaire in the direction of the beam axis.

Photometry: measurement of quantities referring to radiation evaluated according to the visual effect which it produces, as based on certain conventions.

Photopic vision: vision when the eye is adapted to levels of luminance of at least several candela per square metre. Vision mediated essentially or exclusively by cones.

Preheat lamp: hot cathode lamp designed to start with preheating of the electrodes – for example: 'TL'M, 'TL'RS lamp.

Radiation: 1. Emission or transfer of energy in the form of electromagnetic waves or particles.
2. These electromagnetic waves or particles.

Reduced luminance coefficient (r): the product of the luminance coefficient (q) and $\cos^3\gamma$, where γ is the angle of light incidence.

Reflectance (formerly Reflection factor): ratio of the reflected radiant or luminous flux to the incident flux.

Reflection: return of radiation by a surface without change of frequency of the monochromatic components of which the radiation is composed.

Refraction: change in the direction of propagation of radiation determined by change in the velocity of propagation in passing through an optically non-homogeneous medium, or in passing from one medium to another.

Refractor: device in which the phenomenon of refraction is used to alter the spatial distribution of the luminous flux from a source.

Regular reflection: reflection without diffusion in accordance with the laws of optical reflection, as in a mirror.

Retina: membrane at the back of the eye which is sensitive to light stimuli and composed of the photoreceptors (cones and rods) and the nerve cells which transmit the stimulation.

Rod: retinal receptor element which is assumed to be primarily concerned with perception of light stimuli when the eye is adapted to darkness (scotopic vision); rods probably play no part in colour stimulus discrimination.

Scotopic vision: vision when the eye is adapted to levels of luminance below some hundredths of a candela per square metre; the rods are considered to be the principal active elements under these conditions. The spectrum appears uncoloured.

Screen: that part of a luminaire designed to prevent the lamps from being directly visible over a given range of angles.
Note: In practice a screen will also act as a light controller.

Single sided arrangement: an arrangement in which the luminaires are placed on one side only of a carriageway.

Solid angle (Ω): the angle subtended at the centre of a sphere by an area on its surface numerically equal to the square of the radius.
Unit: steradian, sr.

Spacing: the distance, measured parallel to the centre line of the carriageway, between successive luminaires.

Spanwire arrangement: an arrangement in which the luminaires are suspended above the carriageway(s) on transverse wires.

Specific luminaire index (SLI): a quantity which indicates the glare control facility of a luminaire.

Spectral distribution: a. Of a photometric quantity: luminous flux, luminous intensity, etc. The spectral concentration of the photometric quantity as a function of wavelength.
b. Of a radiometric quantity: radiant flux (power), radiant intensity, etc. The spectral concentration of the radiometric quantity as a function of wavelength.
Note: Commonly the relative spectral distribution is used, i.e. the spectral concentration of the photometric or radiometric quantity measured in terms of an arbitrary value of this quantity.

Spectral energy distribution: of a radiation. Description of the spectral character of a radiation by the relative spectral distribution of some radiometric quantity (radiant flux (power), radiant intensity, etc.).

Spectral light distribution: of a radiation. Description of the spectral character of a radiation by the relative spectral distribution of some photometric quantity (luminous flux, luminous intensity, etc.).

Spectral luminous efficiency curve: curve that gives the relative sensitivity (V) of the CIE standard photometric observer for monochromatic radiation in dependence of the wavelength
For photopic vision: $V(\lambda)$ curve
For scotopic vision: $V'(\lambda)$ curve

Specular factor $(S1$ and $S2)$: a factor representing the degree of specular reflection of a road surface.

Specular reflection: see **Regular reflection.**

Specular reflector: that part of a luminaire designed to reflect the luminous flux of the lamps in required directions by means of specular reflection.

Speed of perception: the reciprocal of the minimum exposure time of an object required for it to be detected.

Spread: quantity of a luminaire to indicate the extent to which the light is 'spread out' across the road. Luminaires are classified as being of narrow, average or broad spread.

Staggered arrangement: an arrangement in which the luminaires are placed alternately on either side of the carriageway.

Starter: device for starting a discharge lamp (in particular a fluorescent lamp) that provides for the necessary preheating of the electrodes and/or causes a voltage surge in combination with the series ballast.

Starting device: electrical apparatus that provides the conditions required for starting a discharge.

Stopping distance: see **Driver stopping distance.**

Switch-start fluorescent lamp: fluorescent lamp suitable for operation with a circuit requiring a starter for the preheating of the electrodes, for example: 'TL' standard type.

Thermal radiation: process of emission in which the radiant energy originates in the thermal agitation of the particles of matter (atoms, molecules, ions).

Threshold contrast: the minimum perceptable contrast for a given state of adaptation of the eye.

Threshold zone of a tunnel: first part of a tunnel, which must be lighted in order to enable approaching drivers to perceive (in time for them to react) possible obstacles in this zone.

Throw: characteristic of a luminaire that indicates the extent to which the light is 'thrown' in the lengthwise direction of the road. Luminaires are classified as being of short, intermediate or long throw.

Tilt, angle of: upward inclination of a luminaire from the horizontal.

Transmission: passage of radiation through a medium without change of frequency of the monochromatic components of which the radiation is composed.

Transition zone of a tunnel: the zone between the threshold zone and the interior zone where the luminance may be gradually decreased from the luminance at the end of the threshold zone to the luminance of the interior zone.

Twin-central arrangement: an arrangement in which the luminaires are placed along the central reserve of a dual carriageway on T-shaped masts.

Ultraviolet radiation: radiation for which the wavelengths of the monochromatic components are smaller than those for visible radiation and more than about 1 nm.

Uniformity ratio: a. A measure of the variation of illuminance over a given plane.
b. A measure of the variation of luminance at a given surface; see **Overall uniformity** and **Longitudinal uniformity.**

Upper [lower] (luminous) flux: of a source. The luminous flux emitted above [below] a horizontal plane passing through the source.
Note: It is essential that the plane be specified in every case.

Upward [downward] light output ratio (luminaire efficiency): the ratio of the flux emitted above [below] a horizontal plane passing through the luminaire to the total bare lamp flux.

306

Utilisation factor: ratio of the utilised flux to the luminous flux emitted by the lamps. *Note:* The term 'coefficient of utilisation' is deprecated.

Utilised flux: luminous flux received on the reference surface under consideration.

Veiling luminance: see **Equivalent veiling luminance.**

Visible radiation: any radiation capable of causing a visual sensation direct.

Visual acuity; sharpness of vision:
1. Qualitatively: Capacity for seeing distinctly objects very close together.
2. Quantitatively: Reciprocal of the value (generally in minutes of arc) of the angular separation of two neighbouring objects (points or lines) which the eye can just see as separate.

Visual angle: the angle subtended by an object or detail at the point of observation: it is usually measured in minutes of arc.

Visual comfort: the degree of visual satisfaction produced by the visual environment; for road lighting conditions a moving observer is assumed.

Visual field: of the eye or eyes. The angular extent of the space in which an object can be perceived when the eye(s) regard(s) an object directly ahead. The field may be monocular or binocular.

Visual guidance: the totality of measures taken to give a road user an unambiguous and immediately recognisable picture of the course of the road ahead.

Visual performance: the quantitative assessment of the performance of a task.

Visual reliability of a motorist: the ability of a motorist to continuously select and process, more or less subconsciously, that part of the visual information presented to him that is necessary for the safe control of his vehicle.

Visual system: the group of structures comprising the eye, the optic nerve and certain parts of the brain, which transforms the light stimulus into a complex of nerve excitations, whose subjective correlate is visual perception.

$V(\lambda)$ **curve:** spectral luminous efficiency curve for photopic vision.

$V'(\lambda)$ **curve:** spectral luminous efficiency curve for scotopic vision.

Wavelength (λ): distance in the direction of propagation of a periodic wave between two successive points at which the phase is the same (at the same time).
Unit: metre, m.

Appendix A

Standard Reflection Tables for Dry Road Surfaces (R and N-tables)

ROADSURFACE STANDARD R1   (Q0 TABLE=1.00)

TAN (GAMMA) = R/H-VALUES CORRESPONDING TO THE LISTED NUMBERS OF REFLECTION VALUES

BETA DEG.	TAN	0.0 / 7.50	0.25 / 8.00	0.50 / 8.50	0.75 / 9.00	1.00 / 9.50	1.25 / 10.00	1.50 / 10.50	1.75 / 11.00	2.00 / 11.50	2.50 / 12.00	3.00	3.50	4.00	4.50	5.00	5.50	6.00	6.50	7.00
0	upper	6550	6190	5390	4310	3410	2690	2240	1890	1620	1210	940	810	710	630	570	510	470	430	400
	lower	370	350	330	310	300	290	280	270	260	250									
2	upper	6550	6190	5390	4310	3410	2690	2240	1890	1620	1210	940	800	690	590	520	470	420	380	340
	lower	310	280	250	230	220	200	180	160	150	140									
5	upper	6550	6190	5390	4310	3410	2690	2240	1890	1570	1170	860	660	550	430	360	310	250	220	180
	lower	150	140	120	100	90	80	70	70	60	60									
10	upper	6550	6190	5390	4310	3410	2600	2150	1710	1350	950	660	460	320	240	190	150	120	100	80
	lower	70	60	50	40	40	30	30	30	20	20									
15	upper	6550	6100	5390	4310	3230	2510	1980	1530	1170	790	490	330	230	170	140	110	90	70	60
	lower	50	40	40	30	30	20	20	20	20	20									
20	upper	6550	6100	5390	4310	3230	2420	1800	1390	1080	660	410	280	200	140	120	90	70	60	50
	lower	40	40	30	30	20	20	20	20											
25	upper	6550	6100	5210	4310	3050	2240	1710	1300	990	600	380	250	180	130	100	80	70	50	40
	lower	40	30	30																
30	upper	6550	6100	5210	4310	2960	2070	1620	1210	940	570	360	230	160	120	90	80	60	50	40
35	upper	6550	6100	5210	4310	2870	1980	1530	1170	900	540	340	220	150	120	90	80	60		
40	upper	6550	6100	5210	4310	2870	1890	1480	1120	850	520	330	220	140	110	90	80			
45	upper	6550	6100	5210	3950	2780	1890	1440	1080	850	510	320	210	140	110	90				
60	upper	6550	6100	5030	3860	2690	1800	1440	1030	830	500	310	210	140	110	90				
75	upper	6550	6100	5030	3710	2690	1800	1390	990	840	510	310	220	150	120	90				
90	upper	6550	6100	5030	3710	2690	1800	1390	990	840	520	330	220	170	130	100				
105	upper	6550	6010	5030	3710	2690	1800	1390	1030	860	540	350	240	190	140	110				
120	upper	6550	6010	5030	3710	2690	1890	1440	1080	900	580	380	270	200	140	130				
135	upper	6550	6010	5030	3710	2690	1890	1480	1120	940	610	400	290	220	160	140				
150	upper	6550	6010	5030	3860	2780	1980	1530	1210	990	650	430	310	230	170	150				
165	upper	6550	6010	5030	3950	2780	2070	1620	1300	1030	690	470	340	250	190	160				
180	upper	6550	6010	5030	3950	2780	2240	1800	1390	1110	750	510	380	270	210	160				

ALL VALUES ARE MULTIPLIED BY 10000

ROADSURFACE STANDARD R2   (Q0 TABLE=1.00)

TAN (GAMMA) = R/H-VALUES CORRESPONDING TO THE LISTED NUMBERS OF REFLECTION VALUES

BETA DEG.	0.0 / 7.50	0.25 / 8.00	0.50 / 8.50	0.75 / 9.00	1.00 / 9.50	1.25 / 10.00	1.50 / 10.50	1.75 / 11.00	2.00 / 11.50	2.50 / 12.00	3.00	3.50	4.00	4.50	5.00	5.50	6.00	6.50	7.00
0	5571	5871	5871	5414	4785	4328	3871	3557	3242	2785	2285	2085	1885	1685	1514	1371	1242	1114	1014
	957	900	828	785	742	700	671	628	600	585									
2	5571	5871	5871	5414	4785	4328	3871	3400	3085	2714	2214	1871	1614	1357	1157	985	828	714	614
	542	471	400	357	328	300	257	228	200	185									
5	5571	5871	5871	5414	4785	4171	3714	3242	2785	2085	1642	1242	957	714	542	414	314	242	200
	171	142	128	100	85	85	71	57	57	57									
10	5571	5871	5871	5257	4642	3871	3242	2785	2171	1571	957	585	385	285	200	157	114	85	71
	57	42	42	42	28	28	28	28	28	14									
15	5571	5871	5757	5100	4171	3400	2557	2171	1671	1057	614	357	214	171	114	85	71	57	42
	42	28	28	28	28	14	14	14	14	14									
20	5571	5871	5757	4942	4157	2942	2171	1771	1357	828	471	257	171	128	85	71	57	42	42
	28	28	28	14	14	14	14	14											
25	5571	5871	5485	4642	3714	2628	2014	1514	1142	685	371	214	142	100	71	57	57	42	28
	28	28	28																
30	5571	5871	5414	4328	3400	2171	1700	1300	957	571	300	185	128	100	71	57	42	42	28
35	5571	5871	5285	4014	3085	1857	1542	1114	871	500	257	171	128	85	71	57	42	42	
40	5571	5871	4942	3714	2785	1700	1328	957	742	428	242	157	114	85	71	57			
45	5571	5414	4642	3400	2471	1542	1142	871	642	385	228	157	114	85	57				
60	5571	5257	4328	3085	2171	1428	1085	742	571	342	228	157	114	85	71				
75	5571	5100	4014	2942	2171	1471	1085	771	585	371	242	157	114	85	71				
90	5571	5100	4014	2942	2171	1514	1142	828	642	400	242	157	128	85	71				
105	5571	4942	3871	2942	2171	1542	1200	900	700	428	257	171	142	100	85				
120	5571	4942	3871	2942	2171	1542	1242	957	742	471	300	200	157	114	100				
135	5571	4942	3871	2942	2014	1628	1271	985	771	500	314	214	171	142	128				
150	5571	4785	3714	2942	2014	1628	1300	1014	800	542	342	242	185	171	142				
165	5571	4785	3714	2942	2014	1700	1328	1042	814	571	371	257	214	185	142				
180	5571	4785	3714	2942	2014	1700	1357	1057	828	585	385	300	242	200	157				

ALL VALUES ARE MULTIPLIED BY 10000

ROADSURFACE STANDARD R3   (Q0 TABLE=1.00)

TAN (GAMMA) = R/H-VALUES CORRESPONDING TO THE LISTED NUMBERS OF REFLECTION VALUES

BETA DEG.	0.0 / 7.50	0.25 / 8.00	0.50 / 8.50	0.75 / 9.00	1.00 / 9.50	1.25 / 10.00	1.50 / 10.50	1.75 / 11.00	2.00 / 11.50	2.50 / 12.00	3.00	3.50	4.00	4.50	5.00	5.50	6.00	6.50	7.00
0	4200	4657	4914	5100	5171	5100	5042	4842	4657	4128	3614	3100	2714	2328	2071	1814	1614	1485	1357
	1242	1185	1114	1042	985	928	885	842	800	757									
2	4200	4657	4914	5100	5171	5100	4971	4785	4585	4000	3357	2771	2328	1942	1557	1342	1100	971	857
	757	671	600	542	485	457	414	371	342	314									
5	4200	4585	4842	5042	5028	4971	4657	4328	4000	3171	2328	1742	1285	1042	857	671	514	428	342
	300	242	214	171	142	128	114	100	85	85									
10	4200	4585	4842	4657	4657	4257	3814	3300	2714	1814	1214	857	614	442	342	257	214	157	128
	100	85	71	57	57	42	42	28	28	28									
15	4200	4528	4657	4585	3942	3485	3100	2457	1942	1228	757	500	371	285	228	200	157	114	100
	71	57	57	42	57	28	28	28	28	28									
20	4200	4457	4528	4328	3557	2971	2514	1814	1428	928	542	357	285	228	171	142	128	114	71
	57	57	42	28	28	28	28												
25	4200	4400	4400	4071	3228	2514	2071	1485	1171	771	442	314	228	171	128	114	114	85	57
	57	42	42	28															
30	4200	4400	4257	3814	2914	2200	1671	1271	1014	628	357	271	200	142	114	100	100	57	42
	57																		
35	4200	4328	4128	3485	2585	1942	1428	1128	885	542	328	228	171	128	114	85	71		
40	4200	4257	3942	3171	2257	1685	1228	1000	771	485	285	214	142	114	100	85			
45	4200	4200	3742	2914	2000	1485	1114	885	685	357	257	185	128	100	85				
60	4200	4000	3357	2514	1685	1185	1028	728	557	328	214	142	100	85	57				
75	4200	3871	3100	2257	1485	1042	857	642	485	314	214	128	100	71	57	42			
90	4200	3742	2914	2128	1428	1000	814	628	485	328	200	128	100	71	42				
105	4200	3685	2842	2128	1428	1014	828	642	485	342	214	142	114	71	57				
120	4200	3614	2842	2128	1428	1057	857	657	500	342	214	157	114	85	57				
135	4200	3557	2842	2071	1428	1100	857	642	514	342	228	157	128	100	71				
150	4200	3485	2842	1942	1428	1100	857	642	514	342	228	171	128	114	100				
165	4200	3428	2771	1942	1428	1100	871	657	528	342	242	171	128	114	100				
180	4200	3428	2771	2000	1428	1114	885	671	542	357	242	185	142	128	100				

ALL VALUES ARE MULTIPLIED BY 10000

310

ROADSURFACE STANDARD R4   (Q0 TABLE=1.00)

TAN (GAMMA) = R/H-VALUES CORRESPONDING TO THE LISTED NUMBERS OF REFLECTION VALUES

BETA DEG.	0.0 / 7.50	0.25 / 8.00	0.50 / 8.50	0.75 / 9.00	1.00 / 9.50	1.25 / 10.00	1.50 / 10.50	1.75 / 11.00	2.00 / 11.50	2.50 / 12.00	3.00	3.50	4.00	4.50	5.00	5.50	6.00	6.50	7.00
0	3300 / 2562	3712 / 2412	4125 / 2300	4700 / 2175	4950 / 2112	5037 / 2050	5112 / 1975	5112 / 1912	5112 / 1862	4950 / 1812	4625	4287	3962	3712	3462	3212	3050	2887	2725
2	3300 / 1175	3962 / 1025	4287 / 925	4787 / 825	4950 / 737	5112 / 662	4950 / 612	4950 / 562	4787 / 512	4450 / 462	3800	3387	2975	2637	2312	2012	1750	1525	1325
5	3300 / 325	3962 / 275	4287 / 237	4625 / 200	4950 / 162	4625 / 150	4450 / 137	4287 / 125	3962 / 100	3300 / 100	2637	2062	1650	1325	987	737	575	462	400
10	3300 / 100	3962 / 75	4287 / 62	4375 / 62	4125 / 50	3875 / 50	3550 / 37	3137 / 37	2800 / 37	1900 / 37	1187	787	562	412	300	237	162	137	112
15	3300 / 50	3962 / 50	4125 / 37	4125 / 37	3625 / 37	3137 / 25	2725 / 25	2225 / 25	1812 / 25	1250 / 25	787	500	300	212	162	125	100	75	62
20	3300 / 37	3875 / 37	3875 / 37	3800 / 25	3137 / 25	2637 / 25	2150 / 25	1737 / 25	1325	912	550	325	200	137	100	87	75	62	50
25	3300 / 37	3800 / 25	3712 / 25	3462	2725	2225	1737	1350	1075	687	375	237	162	112	87	75	62	50	37
30	3300	3625	3550	3137	2475	1900	1437	1100	887	562	312	187	137	100	75	62	50	37	37
35	3300	3550	3462	2887	2312	1650	1250	937	737	462	262	162	125	87	75	62	50		
40	3300	3462	3300	2637	2062	1437	1100	825	662	400	212	150	112	87	62	50			
45	3300	3387	3137	2475	1812	1287	987	737	562	350	200	137	100	75	62				
60	3300	3050	2725	2062	1400	962	762	550	412	262	162	125	100	75	62				
75	3300	2887	2475	1737	1075	825	625	462	362	250	150	112	87	75	62				
90	3300	2800	2312	1650	1075	812	625	462	362	250	150	112	87	75	62				
105	3300	2800	2225	1650	1075	812	625	462	362	250	162	112	100	87	75				
120	3300	2725	2150	1562	1075	787	625	475	375	262	162	112	100	87	75				
135	3300	2725	2150	1562	1075	812	650	500	400	275	187	137	112	87	75				
150	3300	2637	2062	1562	1087	825	687	512	412	300	200	150	112	100	75				
165	3300	2637	2062	1487	1087	837	687	525	425	312	212	162	137	112	87				
180	3300	2637	2062	1487	1087	850	687	562	462	325	237	187	150	125	100				

ALL VALUES ARE MULTIPLIED BY 10000

ROADSURFACE STANDARD N1   (Q0 TABLE=1.00)

TAN (GAMMA) = R/H-VALUES CORRESPONDING TO THE LISTED NUMBERS OF REFLECTION VALUES

BETA DEG.	0.0 / 7.50	0.25 / 8.00	0.50 / 8.50	0.75 / 9.00	1.00 / 9.50	1.25 / 10.00	1.50 / 10.50	1.75 / 11.00	2.00 / 11.50	2.50 / 12.00	3.00	3.50	4.00	4.50	5.00	5.50	6.00	6.50	7.00
0	7680 / 230	6940 / 210	5570 / 190	4240 / 180	3230 / 170	2520 / 160	2020 / 150	1640 / 150	1380 / 140	1030 / 140	800	650	550	470	400	350	310	280	250
2	7680 / 170	6940 / 150	5570 / 140	4240 / 130	3220 / 120	2500 / 110	1980 / 100	1620 / 90	1360 / 90	1000 / 80	750	600	480	400	340	280	250	210	190
5	7680 / 80	6940 / 70	5570 / 60	4240 / 50	3210 / 50	2470 / 40	1930 / 40	1540 / 40	1260 / 30	860 / 30	610	450	340	260	200	150	130	110	90
10	7680 / 40	6940 / 40	5550 / 30	4170 / 30	3100 / 20	2340 / 20	1770 / 20	1340 / 20	1040 / 20	640 / 20	410	280	200	140	110	80	70	50	50
15	7680 / 30	6940 / 30	5540 / 20	4150 / 20	3020 / 20	2200 / 20	1600 / 20	1170 / 20	880 / 20	510 / 20	310	210	140	110	80	60	50	40	30
20	7680 / 30	6930 / 20	5500 / 20	4060 / 20	2890 / 20	2060 / 20	1470 / 10	1040 / 10	760	430	260	170	120	80	60	50	40	30	30
25	7680 / 30	6930 / 20	5460 / 20	3970	2780	1930	1350	940	690	380	240	150	110	80	60	50	40	30	30
30	7680	6930 / 20	5440 / 20	3920	2710	1860	1280	890	650	360	220	150	100	80	50	50	40	30	30
35	7680	6930	5440	3880	2660	1800	1240	870	630	350	210	150	100	80	50	50	40	30	30
40	7680	6930	5430	3820	2610	1760	1210	840	610	350	210	150	100	60	50	50	40		
45	7680	6950	5430	3780	2570	1730	1190	840	610	350	210	150	100	80	60	50	40		
60	7680	6990	5420	3810	2590	1750	1220	860	630	370	230	160	120	80	60				
75	7680	7020	5470	3880	2660	1830	1290	930	690	410	260	180	130	100	80				
90	7680	7140	5640	4070	2840	2000	1420	1040	780	480	310	210	160	120	100				
105	7680	7200	5770	4250	3030	2160	1570	1160	880	550	360	250	190	150	120				
120	7680	7340	6000	4500	3280	2370	1750	1310	1010	640	420	310	230	180	140				
135	7680	7410	6150	4690	3460	2540	1890	1440	1110	710	480	350	260	200	180				
150	7680	7510	6330	4890	3680	2710	2040	1550	1210	790	550	390	300	230	180				
165	7680	7530	6400	4970	3750	2790	2120	1620	1270	840	580	420	320	250	210				
180	7680	7570	6460	5050	3810	2850	2160	1660	1310	860	600	440	340	270	210				

ALL VALUES ARE MULTIPLIED BY 10000

312

ROADSURFACE STANDARD N2   (Q0 TABLE=1.00)

TAN (GAMMA) = R/H-VALUES CORRESPONDING TO THE LISTED NUMBERS OF REFLECTION VALUES

BETA DEG.	scale	0.0 / 7.50	0.25 / 8.00	0.50 / 8.50	0.75 / 9.00	1.00 / 9.50	1.25 / 10.00	1.50 / 10.50	1.75 / 11.00	2.00 / 11.50	2.50 / 12.00	3.00	3.50	4.00	4.50	5.00	5.50	6.00	6.50	7.00
0	upper	6771	6743	6100	5343	4657	4057	3557	3129	2771	2243	1857	1571	1343	1143	1000	871	771	686	614
0	lower	557	514	471	429	400	371	357	329	314	300									
2	upper	6771	6729	6086	5314	4614	4014	3514	3086	2714	2143	1714	1400	1143	929	786	657	557	486	429
2	lower	386	329	300	257	243	229	200	186	171	171									
5	upper	6771	6729	6086	5286	4557	3929	3357	2871	2443	1771	1286	929	686	514	400	314	257	200	171
5	lower	143	129	100	86	71	71	57	57	43	43									
10	upper	6771	6714	6029	5157	4329	3586	2900	2314	1829	1171	743	486	329	243	171	129	100	86	71
10	lower	57	43	43	29	29	29	14	14	14	14									
15	upper	6771	6714	5971	4986	4057	3200	2457	1843	1386	814	500	314	214	143	114	86	71	57	43
15	lower	29	29	29	14	14	14	14	14	14										
20	upper	6771	6686	5900	4800	3757	2829	2086	1514	1114	629	371	229	157	114	86	57	57	43	29
20	lower	29	29	14	14	14	0	0												
25	upper	6771	6657	5829	4586	3471	2514	1814	1300	929	529	314	200	129	100	71	57	43	29	29
25	lower	29	14	14																
30	upper	6771	6629	5700	4414	3243	2300	1614	1157	814	457	271	186	129	86	71	43	43	29	
35	upper	6771	6614	5571	4243	3043	2129	1486	1043	757	429	257	171	114	86	57	43	43		
40	upper	6771	6557	5457	4071	2886	1986	1386	971	686	400	243	157	114	86	57	43			
45	upper	6771	6429	5357	3929	2757	1886	1314	914	657	386	243	157	114	86	57				
60	upper	6771	6429	5129	3643	2529	1729	1200	857	629	371	229	157	114	86	57				
75	upper	6771	6329	4957	3500	2429	1671	1186	843	629	371	243	157	114	86	71				
90	upper	6771	6271	4900	3486	2429	1700	1214	886	657	400	257	171	129	86	71				
105	upper	6771	4929	4871	3500	2486	1757	1271	929	700	429	286	200	143	114	86				
120	upper	6771	6200	4914	3614	2571	1843	1343	1000	757	471	314	229	171	129	100				
135	upper	6771	6200	4971	3657	2657	1929	1414	1057	814	529	343	257	186	143	114				
150	upper	6771	6214	5029	3743	2743	5914	1500	1129	871	571	386	271	214	157	129				
165	upper	6771	6229	5043	3786	2800	2071	1543	1171	914	600	400	300	229	171	143				
180	upper	6771	6229	5086	3829	2843	2100	1571	1200	929	614	414	300	229	186	143				

ALL VALUES ARE MULTIPLIED BY 10000

313

ROADSURFACE STANDARD N3   (Q0 TABLE=1.00)

TAN (GAMMA) = R/H-VALUES CORRESPONDING TO THE LISTED NUMBERS OF REFLECTION VALUES

BETA DEG.	0.0 / 7.50	0.25 / 8.00	0.50 / 8.50	0.75 / 9.00	1.00 / 9.50	1.25 / 10.00	1.50 / 10.50	1.75 / 11.00	2.00 / 11.50	2.50 / 12.00	3.00	3.50	4.00	4.50	5.00	5.50	6.00	6.50	7.00
0	5057	5586	5800	5786	5657	5471	5143	4786	4457	3857	3329	2886	2529	2214	1957	1729	1543	1386	1271
	1157	1057	971	900	829	771	729	686	643	614									
2	5057	5586	5771	5757	5600	5343	5014	4643	4257	3571	2943	2443	2029	1686	1429	1200	1029	886	786
	700	614	543	486	443	400	371	343	314	300									
5	5057	5571	5757	5700	5486	5143	4686	4200	3700	2843	2114	1557	1171	871	671	529	414	343	286
	243	214	171	157	129	114	100	86	86	71									
10	5057	5557	5671	5486	5100	4529	3857	3200	2600	1714	1114	729	500	343	257	186	157	129	100
	86	71	57	57	43	43	43	29	29	29									
15	5057	5543	5571	5214	4600	3843	3071	2357	1800	1086	657	414	286	200	143	114	86	71	57
	57	43	43	43	29	29	14	14	14	14									
20	5057	5500	5414	4886	4100	3243	2457	1814	1343	757	443	286	186	143	100	86	71	57	43
	43	29	29	29	29	29													
25	5057	5443	5229	4557	3629	2757	2014	1457	1071	614	371	243	157	114	86	71	57	57	43
	43	29	29	29	29														
30	5057	5400	5071	4257	3257	2400	1700	1229	900	500	300	200	143	100	86	57	57	43	29
35	5057	5329	4900	3971	2957	2129	1500	1071	800	457	271	186	129	100	71	57	57		
40	5057	5271	4714	3714	2700	1914	1343	957	714	400	257	171	129	86	71	57			
45	5057	5214	4543	3486	2500	1757	1229	871	643	371	243	157	114	86	71				
60	5057	5014	4157	3043	2100	1471	1029	743	543	329	214	143	100	71					
75	5057	4829	3829	2743	1900	1329	943	671	500	314	200	143	100						
90	5057	4686	3657	2600	1800	1271	900	657	500	300	200	143	100						
105	5057	4557	3529	2514	1771	1243	900	671	500	314	214	157	114						
120	5057	4471	3457	2500	1757	1271	914	686	529	329	229	157	114						
135	5057	4400	3429	2500	1786	1300	957	714	543	343	243	171	129						
150	5057	4371	3429	2514	1814	1343	986	743	571	386	257	186	143						
165	5057	4343	3429	2529	1843	1357	1014	771	600	400	271	200	157						
180	5057	4343	3429	2529	1857	1371	1029	786	614	400	286	214	157						

ALL VALUES ARE MULTIPLIED BY 10000

314

ROADSURFACE STANDARD N4   (QO TABLE=1.00)

TAN (GAMMA) = R/H-VALUES CORRESPONDING TO THE LISTED NUMBERS OF REFLECTION VALUES

BETA DEG.	0.0	0.25	0.50	0.75	1.00	1.25	1.50	1.75	2.00	2.50	3.00	3.50	4.00	4.50	5.00	5.50	6.00	6.50	7.00	7.50	8.00	8.50	9.00	9.50	10.00	10.50	11.00	11.50	12.00
0	3525	4150	4688	5150	5513	5738	5825	5800	5663	5313	4813	4363	3950	3575	3250	2963	2713	2475	2300	2113	1975	1850	1725	1638	1575	1475	1375	1288	1225
2	3525	4150	4675	5138	5475	5663	5700	5613	5388	4838	4163	3575	3063	2588	2188	1838	2025	1350	1188	1063	938	850	763	688	625	575	525	488	450
5	3525	4138	4663	5075	5375	5375	5213	4900	4450	3538	2638	1975	1463	1100	863	675	538	438	350	288	250	213	175	150	138	125	100	88	75
10	3525	4100	4588	4850	4813	4563	4075	3550	2925	1900	1225	825	588	413	300	238	188	150	113	100	88	75	63	50	38	38	25		
15	3525	4100	4475	4513	4263	3738	3075	2488	1913	1163	713	463	338	250	188	150	113	88	75	63	50	50	50	38	25	25			
20	3525	4050	4313	4213	3675	3038	2363	1825	1350	800	475	313	225	175	125	100	75	63	63	50	38	38	38	38					
25	3525	3988	4138	3925	3138	2475	1863	1425	1063	625	375	250	175	138	100	75	63	63	63	38	38	38	38						
30	3525	3938	3975	3550	2750	2100	1525	1150	863	513	313	213	150	113	88	75	63	63	50	38	38	38							
35	3525	3888	3813	3200	2450	1825	1313	988	750	450	275	188	138	100	75	63	50	50	25										
40	3525	3788	3588	2938	2175	1600	1150	863	650	388	238	163	125	88	63	63													
45	3525	3688	3388	2725	1950	1438	1038	788	575	350	213	150	113	88	63														
60	3525	3500	3025	2275	1575	1138	838	625	463	288	175	125	88	75	63														
75	3525	3275	2613	1913	1325	950	700	525	400	250	163	113	75	75	63														
90	3525	3163	2450	1775	1213	888	663	500	375	238	150	100	75	63	50														
105	3525	3088	2363	1725	1175	850	650	488	375	238	150	113	75	63	50														
120	3525	3013	2300	1688	1175	863	650	500	375	238	163	113	88	75	63														
135	3525	2963	2250	1650	1163	863	663	513	388	250	175	125	100	75	63														
150	3525	2888	2213	1625	1175	875	675	525	400	263	175	125	100	88	63														
165	3525	2875	2200	1625	1175	888	688	538	413	288	188	150	113	88	75														
180	3525	2850	2188	1625	1188	900	688	550	425	288	200	150	113	100	75														

ALL VALUES ARE MULTIPLIED BY 10000

315

Appendix B

Standard Reflection tables for Wet Road Surfaces (W-tables)

WET ROADSURFACE STANDARD W1 (Q0 TABLE=1.00 CORRESPONDING Q0 DRY=0.77)

TAN (GAMMA) = R/H-VALUES CORRESPONDING TO THE LISTED NUMBERS OF REFLECTION VALUES

BETA DEG.	0.0	0.25	0.50	0.75	1.00	1.25	1.50	1.75	2.00	2.50	3.00	3.50	4.00	4.50	5.00	5.50	6.00	6.50	7.00
0	3456	3736	4394	5683	7306	8823	9981	10683	10893	10385	9271	8017	6754	5499	4456	3631	2956	2438	2052
2	3456	3736	4377	5622	7175	8560	9508	10008	10025	9139	7683	6227	4806	3605	2763	2105	1640	1298	1044
5	3456	3736	4333	5455	6780	7736	8157	8201	7692	6104	4324	2982	2017	1368	956	658	474	342	272
10	3456	3728	4219	5052	5710	5929	5561	4859	4035	2570	1561	921	570	351	228	149	105	79	61
15	3456	3710	4026	4491	4587	4245	3526	2780	2131	1210	649	360	193	123	79	53	44	35	26
20	3456	3666	3815	3894	3605	3008	2324	1702	1245	658	351	202	114	79	53	44	35	26	18
25	3456	3622	3614	3377	2833	2123	1535	1044	728	360	193	114	70	53	35	35	26	18	18
30	3456	3587	3412	2973	2298	1640	1131	763	526	263	149	88	61	44	35	26	18	18	18
35	3456	3552	3219	2614	1859	1272	833	553	377	193	114	70	53	35	26	26	18		
40	3456	3491	3035	2342	1605	1061	693	465	325	167	96	61	44	35	26	18			
45	3456	3429	2859	2096	1386	895	579	395	272	149	88	61	44	35	26				
60	3456	3307	2561	1737	1123	737	491	342	246	140	88	61	44	35	26				
75	3456	3184	2386	1605	1061	710	491	342	254	149	96	70	53	35	35				
90	3456	3114	2333	1605	1088	746	526	386	281	175	114	79	61	44	35				
105	3456	3079	2315	1640	1140	798	570	421	316	193	132	96	70	53	44				
120	3456	3070	2368	1710	1210	860	631	465	360	228	149	105	79	61	53				
135	3456	3052	2394	1754	1272	921	675	509	395	246	167	123	96	70	61				
150	3456	3070	2447	1833	1333	982	728	553	421	272	193	140	105	79	61				
165	3456	3070	2473	1859	1368	1009	754	570	447	298	202	149	114	88	70				
180	3456	3087	2491	1894	1395	1026	772	588	465	307	210	158	114	96	79				

Continuation (TAN GAMMA 7.50–12.00, under columns 0.0–2.50):

BETA DEG.	7.50	8.00	8.50	9.00	9.50	10.00	10.50	11.00	11.50	12.00
0	1745	1491	1272	1105	974	860	763	684	623	570
2	860	702	579	491	421	368	325	298	263	246
5	210	167	132	114	96	79	61	61	53	44
10	53	35	35	26	18	18	18	18	18	18
15	26	18	18	18	18	18	18	9	18	9
20	18	18	18	9	9					
25	18	18								

ALL VALUES ARE MULTIPLIED BY 10000

WET ROADSURFACE STANDARD W2 (Q0 TABLE=1.00 CORRESPONDING Q0 DRY=0.61)

R/H-VALUES CORRESPONDING TO THE LISTED NUMBERS OF REFLECTION VALUES

BETA DEG.	TAN (GAMMA) 0.0 / 7.50	0.25 / 8.00	0.50 / 8.50	0.75 / 9.00	1.00 / 9.50	1.25 / 10.00	1.50 / 10.50	1.75 / 11.00	2.00 / 11.50	2.50 / 12.00	3.00	3.50	4.00	4.50	5.00	5.50	6.00	6.50	7.00
0	2273	2439	2891	4127	5995	8295	10342	11950	13007	13758	13313	12077	10482	8753	7198	5889	4898	4101	3423
	2871	2439	2074	1775	1529	1316	1150	1017	904	818									
2	2273	2433	2871	4108	5989	8149	9943	11226	11911	11711	10568	8906	7225	5583	4287	3250	2539	2001	1589
	1269	1030	831	685	578	485	419	366	326	292									
5	2273	2433	2838	3995	5636	7378	8361	8793	8547	7211	5357	3755	2546	1615	1070	731	512	366	266
	206	160	120	93	73	66	53	47	40	33									
10	2273	2413	2772	3649	4699	5244	5231	4633	3835	2499	1436	831	472	266	166	106	73	53	40
	27	27	20	20	13	13	13	13	7	7									
15	2273	2406	2665	3184	3589	3529	2951	2340	1728	851	432	226	133	80	53	33	27	20	20
	13	13	13	13	7	7	7	7	7	7									
20	2273	2373	2519	2718	2645	2273	1715	1263	877	432	213	120	73	47	33	27	20	13	13
	13	13	7	7	7	7	7	7											
25	2273	2346	2379	2313	1947	1462	990	678	445	219	106	60	40	27	20	20	13	13	13
	13	7	7																
30	2273	2320	2233	1994	1529	1057	705	472	312	160	86	47	33	20	20	13	13	13	7
35	2273	2293	2094	1715	1196	764	498	326	226	120	66	40	27	20	13	13	13		
40	2273	2260	1987	1529	1017	651	425	279	193	106	60	33	27	20	13	7			
45	2273	2220	1881	1363	864	552	359	239	173	93	60	33	27	20	13				
60	2273	2140	1682	1137	731	479	319	226	160	93	60	40	27	20	20				
75	2273	2060	1549	1070	711	472	326	233	173	100	66	40	33	27	20				
90	2273	2034	1542	1077	738	505	352	259	186	113	73	53	40	27	20				
105	2273	2007	1549	1097	764	532	379	279	206	126	86	60	40	33	27				
120	2273	2021	1582	1143	811	572	419	306	239	146	100	73	53	40	33				
135	2273	2027	1602	1176	844	605	445	332	253	160	106	80	60	47	33				
150	2273	2047	1642	1230	891	651	479	359	279	179	120	86	66	53	40				
165	2273	2040	1648	1243	911	665	492	372	292	186	126	93	73	60	47				
180	2273	2047	1668	1263	931	678	512	385	299	193	133	100	73	60	47				

ALL VALUES ARE MULTIPLIED BY 10000

317

WET ROADSURFACE STANDARD W3 (Q0 TABLE=1.00 CORRESPONDING Q0 DRY=0.49)

TAN (GAMMA) = R/H-VALUES CORRESPONDING TO THE LISTED NUMBERS OF REFLECTION VALUES

First value row of each BETA corresponds to TAN = 0.0 … 7.00. The second (smaller) value row corresponds to TAN = 7.50 … 12.00 placed under the first ten columns.

BETA DEG	0.0 / 7.50	0.25 / 8.00	0.50 / 8.50	0.75 / 9.00	1.00 / 9.50	1.25 / 10.00	1.50 / 10.50	1.75 / 11.00	2.00 / 11.50	2.50 / 12.00	3.00	3.50	4.00	4.50	5.00	5.50	6.00	6.50	7.00
0	1576	1658	2046	3183	5071	7525	9871	11938	13606	15550	15708	14830	13458	11535	9805	8076	6709	5612	4785
	4071	3479	2969	2551	2183	1888	1648	1449	1280	1153									
2	1576	1663	2051	3173	4979	7336	9387	11402	12427	13075	12152	10382	8606	6821	5270	4040	3209	2571	2076
	1663	1357	1102	898	750	633	541	464	413	367									
5	1576	1658	2025	3051	4714	6489	7805	8525	8520	7035	5285	3699	2510	1597	1036	673	469	337	245
	184	143	107	87	66	56	51	41	36	31									
10	1576	1653	1959	2729	3836	4596	4561	4035	3362	1990	1122	592	332	184	112	71	51	36	31
	20	20	15	15	10	10	10	10	10	10									
15	1576	1643	1888	2362	2780	2734	2255	1673	1158	556	265	143	71	46	31	20	15	15	10
	10	10	10	10	5	5	5	5	5	5									
20	1576	1622	1755	1964	1918	1612	1153	801	520	245	117	71	41	31	20	15	15	10	10
	10	10	10	10	5	5	5	5	5										
25	1576	1602	1632	1638	1326	949	587	383	235	107	51	36	26	20	15	10	10	10	10
30	1576	1587	1530	1367	1025	679	418	270	173	87	51	31	20	15	10	10	10	10	
35	1576	1566	1434	1143	791	490	301	194	133	71	46	26	20	15	10	10	10		
40	1576	1551	1357	1020	679	423	265	179	122	71	46	31	20	15	10	10			
45	1576	1530	1280	913	582	372	240	158	117	66	41	31	20	15	10	10			
60	1576	1485	1153	801	520	347	235	163	117	71	46	31	20	15	10				
75	1576	1449	1107	770	515	347	240	173	128	77	46	31	26	20	15				
90	1576	1439	1107	781	536	372	260	189	138	82	56	41	31	20	15				
105	1576	1428	1107	796	556	388	281	204	153	92	61	46	31	26	20				
120	1576	1439	1133	826	587	418	301	224	168	102	71	51	36	31	20				
135	1576	1434	1143	847	607	434	316	235	184	117	77	56	41	31	26				
150	1576	1444	1173	877	638	464	342	255	199	128	87	61	46	36	31				
165	1576	1444	1173	888	648	474	352	265	204	133	92	66	51	41	31				
180	1576	1449	1184	903	658	485	362	270	214	138	97	71	51	41	36				

ALL VALUES ARE MULTIPLIED BY 10000

318

WET ROADSURFACE STANDARD W4 (Q0 TABLE=1.00 CORRESPONDING Q0 DRY=0.42)

TAN (GAMMA) = R/H-VALUES CORRESPONDING TO THE LISTED NUMBERS OF REFLECTION VALUES

BETA DEG.	0.0 / 7.50	0.25 / 8.00	0.50 / 8.50	0.75 / 9.00	1.00 / 9.50	1.25 / 10.00	1.50 / 10.50	1.75 / 11.00	2.00 / 11.50	2.50 / 12.00	3.00	3.50	4.00	4.50	5.00	5.50	6.00	6.50	7.00
0	1149	1117	1206	1777	3262	5545	8103	10523	12458	15785	16991	17084	16639	15028	13296	11495	10111	8949	7630
0 (7.50–12.00)	6601	5658	4812	4145															
2	1149	1117	1218	1769	3193	5262	7399	9212	10799	12223	12045	10969	9621	7852	6306	4978	3910	3088	2424
2 (7.50–12.00)	1906	1506	1214	1008	850	712	611	530	465	417									
5	1149	1117	1206	1769	2971	4327	5379	6144	6472	5525	4270	2926	2028	1311	765	506	304	206	142
5 (7.50–12.00)	101	81	61	49	40	32	28	24	20	20									
10	1149	1121	1170	1692	2339	2777	2789	2615	2117	1113	611	291	146	77	49	28	20	16	16
10 (7.50–12.00)	12	12	8	8	8	8	8	8	8	8									
15	1149	1129	1154	1429	1623	1457	1137	854	652	267	105	57	32	24	16	12	12	8	8
15 (7.50–12.00)	8	8	8	4	4	4	4												
20	1149	1125	1089	1145	1077	830	583	405	279	134	61	32	20	16	12	12	12	8	8
20 (7.50–12.00)	8	8	8	4	4	4													
25	1149	1117	1028	915	712	474	295	190	121	69	36	20	16	12	8	12	8	8	8
25 (7.50–12.00)	8	8																	
30	1149	1113	988	809	583	376	243	158	109	61	32	20	12	12	8	8	8	8	8
35	1149	1113	947	712	474	300	198	134	97	53	32	16	12	12	8	8	8		
40	1149	1105	915	672	437	279	186	125	89	49	32	20	12	12	8	8			
45	1149	1097	882	631	405	263	174	121	85	49	32	24	16	12	12				
60	1149	1085	850	587	385	259	174	121	89	53	32	24	16	12	12				
75	1149	1064	818	575	389	263	182	125	93	53	36	24	16	12	12				
90	1149	1060	826	587	405	275	198	142	105	65	40	28	20	16	12				
105	1149	1056	830	599	417	291	206	150	113	69	45	32	24	20	16				
120	1149	1069	854	619	441	312	223	166	125	77	53	36	28	24	20				
135	1149	1073	862	635	449	324	235	174	134	85	57	40	32	24	20				
150	1149	1077	878	656	478	344	251	190	146	93	65	49	36	28	24				
165	1149	1081	886	668	486	352	259	194	150	97	69	49	36	28	24				
180	1149	1081	890	676	494	364	267	202	158	101	73	53	40	32	28				

ALL VALUES ARE MULTIPLIED BY 10000

Appendix C

Glare Nomograms

Threshold Increment *(TI)*

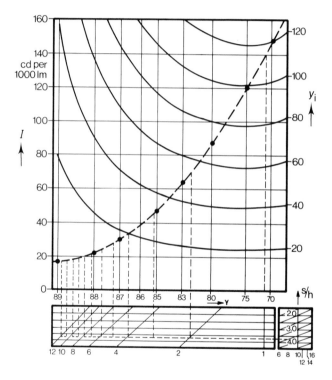

The above nomogram yields the veiling luminance L_v produced on the eye of an observer in line with a row of twelve luminaires and located a distance off from the first luminaire such that this just appears within the standard screening angle of 20°. Knowing L_v, the threshold increment TI can be calculated using

$$TI = 65 \frac{L_v}{L_{av}^{0.8}}$$ where L_{av} is the average road surface luminance.

The procedure is as follows:
First, in the upper diagram, plot the luminous intensity I (in cd/1000 lm) as a function of the angle of emission γ (measured from the downward vertical) for the luminaire in use in the $C = 0°$ plane (as illustrated by the dotted curve shown in the example above).

320

Next, in the lower diagram, draw a horizontal line (shown dotted in the example given) from the point on the s/h scale corresponding to the spacing/mounting-height ratio in use. Project the points where this line intersects the single vertical line and the eleven sloping lines in this diagram vertically upwards to cut the curve previously constructed in the upper diagram. With the twelve Y_i values read off from the scale on the right of this diagram calculate L_v using

$$L_v = \frac{2.8 \times 10^{-3}}{(h - 1.5)^2} \Phi_1 \sum_{i=1}^{12} Y_i$$

where
$h =$ luminaire mounting height (m)
$\Phi_1 =$ luminous flux of bare lamp (lm)
$Y_i =$ values read off from upper diagram

In the example illustrated, where $\Phi_1 = 25000$ lm, $s = 37$ m and $h = 10$ m:

Luminaire No.	Y_i
1	115
2	53
3	31
4	19
5	15
6	12
7	9
8	8
9	7
10	6
11	5
12	4

$$\Sigma Y_i = 284$$

From which
$$L_v = \frac{2.8 \times 10^{-3}}{(10 - 1.5)^2} 25000 \times 284$$

i.e. $L_v = 0.28$ cd/m^2

Glare Control Mark (*G*)

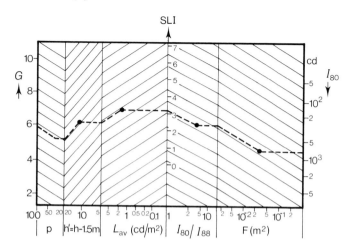

The above nomogram can be used in place of the glare formula given in Chapter 13 to find the glare control mark of a road lighting installation. Proceed as follows:

Start at the right-hand side of the diagram and find the value of I_{80} for the luminaire in use. Draw a horizontal line from this point to a point vertically above the value on scale *F* corresponding to the flashed area of the luminaire. From here, follow the slope of the lines in this area of the diagram until the second area is entered. Work from point to point in this way until the glare scale on the left-hand side of the diagram is reached, and read off the (uncorrected) glare mark.

If the specific luminaire index (*SLI*) of the luminaire is known, start at the middle of the diagram and proceed as above.

The value of *G* obtained from the nomogram must be corrected according to the colour factor of the lamp type in use.

In the example illustrated:

$$I_{80} = 790 \text{ cd}$$
$$F = 0.03 \text{ m}^2$$
$$I_{80}/I_{88} = 3$$
$$L_{av} = 2 \text{ cd/m}^2$$
$$h' = 13.5 \text{ m}$$
$$p = 30 \text{ (number of luminaires per km)}$$
$$SLI = 3.2$$

Giving *G* (uncorrected) = 5.9 and
G (corrected according to CIE for SOX lighting) = 6.3

Index

luxmeter 212, 303